HUMAN ORIGINS: LANGUAGE, SYMBOLISM, AND SECRETS OF OUR EVOLUTION

Author: Rebbie Monkie, **MSc**

Contents:

Preface

This book explores the origins of human beings using both the theory of evolution, and sociological understandings and experiences to uncover what makes us special in the animal kingdom. Specifically, if we are all, for want of a better word, 'psychotic', from a certain perspective. We are the only animal that thinks and communicates mainly using symbols, and I'm arguing that we are all mad due to this symbol use, since mad people see and hear things that aren't there similarly to how all extant peoples think, express, and communicate things that sometimes have no basis in reality (through the use of symbols). For example, a ghost is not a real thing, but part of a myth based around the dead and understanding what happens when we die. There's no physical evidence for ghosts, yet we all believe in them as a concept at least. The book is accessible, and anyone can read it and wonder about human origins. I have a background in anthropology, bioarchaeology, and evolutionary psychology. I'm educated to master's level.

The book has 9 chapters. Each are quite short and to the point. I'm trying to argue for the versatility and freedom of expression that the evolution of symbolism has facilitated in our species. I discuss human evolution generally in chapter 1, and 3. Chapter 2 and 4 cover similarities to animals in our cognition and intelligence. While chapter 6 and 7 explore the evolution of language and symbolism. Chapter 5 covers whether industrialised western society is detrimental to mental health, and how governments might capitalise on this vulnerability. Chapter 8 looks at race, discrimination, and how labelling can divide rather than unite. Finally, chapter 9 looks at the future of humanity, and whether we can save ourselves from extinction.

CHAPTER 1: ALMOST HUMAN

To actually define what being human means in relation to the fossil record has proven not to be as easy as one would first think. The fine line that exists between one animal species and a closely related similar animal species is often moved depending on how you define species (Aldhebiani, 2017). The same applies for genus and other types of taxa (especially when a new organism is found which contradicts previous definitions). When we say human, do we mean our species, or our genus? I think we usually mean our species, but then, would a Neanderthal have really been that different to us? In the modern era we are obsessed with discontinuous categories and like to label things and call them special. This is sometimes relevant, such as the difference between *Australopithecus* and *Homo sapiens*, where a comparison would show vast phenotypic differences.

When we talk about human evolution, fossils are so few and far between that in many cases the comparisons are based on single examples (Stringer and Barnes, 2015), and even where there are many, we often don't know how different (or similar) these groups were socially or culturally. However, many of the species in our genus that lived before ours were intelligent, innovative, and resourceful. Neanderthals on average had a bigger brain capacity than ourselves (Pearce *et al*,2013), yet for some reason we seem to think both that they were less intelligent than us, and also that brain capacity is the main thing that matters in terms of intelligence, the answers to both these questions I'm very doubtful about.

Early Neanderthals, one of our most recent extinct ancestors, were more robust than modern humans (Dannemann and Kelso, 2017), had heavy brow ridges and used animal skins as clothes for warmth, similar to ourselves (Wales, 2012). They were highly socially complex, and cooperated in big game hunts that must have required great bravery and planning (Marin and Carbonell, 2017) similarly to hunter-gatherers today. Interestingly often their teeth were highly worn down. It is thought their teeth were highly worn because they used them as tools in and of themselves (Clement *et al*, 2012). It is thought they would use them like a vice to hold animal hides while preparing them and that their jaws were used as tools more consistently and often than is the case with modern humans (Clement *et al*, 2012). This implies a cultural tradition, as the skill was probably passed down through the generations, which qualifies as a cultural tradition. However, there is not much symbolic culture associated with early Neanderthals. Evidence of art, music, and early religion or spiritual belief (Mellars, 2010) are rare among the material culture that we have from Neanderthals. Stylistic expression can also be classified as symbolic culture, whether that be etchings of tools, or etchings on bones after having processed a carcass for eating. This type of evidence seems to be more common later in the archaeological record. What seems to have coincided with early stylistic expression is a diversification of technological creations, such as needles and spear throwers (d'Errico *et al*, 2018). It is only with early modern humans around 150,000 years ago, primarily in Africa, that we start to see this early evidence of symbolism (Henshilwood, 2009). Neanderthals hunted large game and cared for their sick and old, yet there's no evidence of vibrant culture, it makes one think, how did these peoples cooperate and communicate with no language or spiritual beliefs?

What was this symbolism that I'm talking about based on? Symbolic culture uses special signs, called symbols. From this, the whole of human culture has developed. Symbols are given meaning through consensus. We are symbolic creatures that think with symbols (Wengrow and Graeber, 2015). Language is a communication system based on symbols (Pierarinen, 2012). But, before we were communicating in symbols we were probably thinking in symbols (Grouchy *et al*, 2016). This book will argue that symbolism is not that different from psychosis, and that it is different only on the level of metaphorical code that it uses. Symbols and metaphor are similar as both are arbitrary (arbitrary in that they don't have to be visually similar to what they represent). Psychotic people often see patterns and structures (visually and based on sound) in things where a 'sane' person wouldn't interpret them as being structured. They could believe furniture, and objects in the environment are displaying emotion, or feeling. If you've ever been really ill with a fever that clouds your mental clarity, maybe you could relate to this kind of feeling. Abstract thinking is a defining feature of symbolism and is not too dissimilar to this.

The early part of human evolution (since the split from chimpanzees at our last common ancestor) didn't have many great innovations (Corvinus, 2004) that we know about. For example, the hand axe, created by early *Homo erectus* around 2 million years ago didn't change a great deal for 1 million years (Corvinus, 2004). However, from *Homo erectus* onwards people were hunting game (Ben-Dor *et al*, 2011), and this requires planning and cooperation. Language would have greatly helped this planning and cooperation. So, was a simple form of language being used? It's widely accepted that symbolism didn't emerge until late in the human story, but absence of evidence isn't evidence of absence. One problem is we currently only have a few fossils and hardly any material culture from the early point of human evolution. However, not all evidence has to be physical.

Symbolism in the early human lineage includes the early emergence of music (Savage *et al*, 2020). Researchers such as Professor Ed Hagen argue that fully formed human music, which is a universal among modern human groups, has it's origins in animal communication (Mehr *et al*, 2021). We can see that bird song, or whale communication is quite similar to human music, and gibbons an ape species closely related to us, produce amazing highly coordinated duets which also closely resemble human music (Milo, 2022). So, it is possible that early humans produced some form of music. However, what were the potential evolutionary explanations for music. If human music is a part of symbolic culture which is exclusive to modern humans, and nothing in the animal kingdom is analogous to it, then how can it have emerged on a continuum with animal music? Perhaps, symbolism hijacked early human communication systems and repurposed them for use in symbolic ritual (Milo, 2022).

A big boundary in human evolution is with the birth of the *Homo* genus which is usually attributed to the emergence of *Homo habilis* (Dunsworth, 2010). These populations are often considered to be the start of the genus because of the comparatively higher proportion of meat in the diet compared to those of the genus *Australopithecus* (Dunsworth, 2010). These hominins are thought to have lived in savannah environments scavenging carcasses and may have filled similar ecological niches to vultures or even hyenas (Bickerton *et al*, 2011).

However, in terms of anatomy and form being similar to humans of today, this only develops in *Homo erectus* which, as well as being relatively more gracile and tall, they also have a bigger brain capacity. *Homo erectus* is also associated with the creation and management of fire (Wrangham, 2017). Control of fire is an amazing feat, especially in comparison to other creatures, none of whom, as far as we know, have achieved this. There is ongoing debate about when control of fire began, but the shortening of the gut in *Homo erectus* (Anton and Snodgrass, 2012) implies that the food being eaten didn't need as much processing as in animals with long guts so it is probable that both more meat was being eaten (which is more easily digestible than starch plant foods) and also food was being cooked (which breaks down food similarly to what happens in the gut). Fire would also have scared away predators. From modern ethnographic studies of hunter-gatherers, and associated papers (Wiessner *et* al, 2014) we know that sitting by a central fire is a social experience and it's where stories are told, and dancing and eating is done. It is, and was, basically party central. The same activities may have occurred with early ancestors like *Homo erectus*. However, what were the issues of the day and how were they communicated? This is still largely a mystery.

The beginnings of the genus are where many researchers begin to call these populations human (Henke, 2011). With *Homo erectus* a great increase in brain capacity emerged, from 800cc in *Homo habilis* to 1000cc or more in *erectus* (Shultz *et al*, 2012). This, correspondingly may have been parallel to increasingly complex behaviour, such as care for the sick or injured (Spikins *et al*, 2022), hunting that requires planning (Voormolen, 2008), and complex manufacturing of lithic tools (of which some may have been composite) (Rocca *et al*, 2017). The evidence we have of these care-giving behaviours include a skull from 1.7 million years ago that was toothless, but the sockets had healed over long before death, meaning someone would have had to have broken up or chewed this individuals' food, or in other words cared for them (Lordkipanidze *et al*, 2005). In animals, care-giving to the sick/injured is so rare, that even a sprain of an ankle, and more so a bone break, usually will lead to death in that animal whether from starvation or predation. This wasn't the case with early humans as care to the sick/injured goes back far into the hominin line (Spikins *et al*, 2022). We know this from signs of recovery from serious injury on bone in human remains we have found which implies care (Spikins, 2018).

A few butchery sites including one from Olduvai Gorge, Tanzania (Yravedra, 2021) where a carcass showed cut marks from lithic tools show that *Homo ergaster/erectus* were not only scavenging, but probably hunting large game. In the Olduvai Gorge site, they as far as we know had primary access to the carcass which implies, they were the group that killed the animal (Yravedra, 2021). Usually if human cut marks are beneath the marks left from scavenging by other animals it means the humans were the ones who killed it, and had primary access to it (with there being the possibility the human hunters had successfully defended it from large carnivores such as Wolves, Lions, or Hyenas).

It's widely known that the handaxe was created by *erectus* (*ergaster*), and they continued to use this tool almost unchanged for a million years (Hicks, 2014). This slow speed of progress is quite difficult to understand when comparing it to that of the last few hundred years, or even last 70,000 if we include the emergence of new art forms such as musical performances, storytelling and technological innovations such as bone needles or tailored clothes (Hoffecker, 2005). Because of this slow progression it is an argument against early hominins having had full symbolism, or more accurately fully-fledged cumulative culture, where cultural and technological advancements accumulate through the generations possibly aided by symbolic communication like language.

After this period, moving on many thousands of years, we get to the era of *Homo heidelbergensis* of which, an even greater rate of encephalisation can be found (Rightmire, 2004). This large brain size is associated with more complex stone (lithic) tools as well as surviving evidence of wooden spears used for hunting (Schoch *et al*, 2015). The slow progress of the Lower Palaeolithic (Early Stone Age in Africa) is now over, and technological advancements in terms of stone tools are beginning to speed up with multiple different lithic cultures, or stone exploitation techniques emerging around the Middle Palaeolithic (Middle Stone Age) time period 300,000 years ago (Minet *et al*, 2021). These individuals were also robust and tall, being comparable to a modern day rugby player (Harvati, 2007). What is most interesting about populations that fall into this category (according to modern taxonomy, of which the lines can be fine and there's often arguments about where the line dividing two species or subspecies actually rests) is how there is much stronger and more consistent evidence this species hunted large game for food (Voormolen, 2008).

The site of Schoningen gives us the impression that many horses, strong wild horses, were killed and butchered here after being cornered by the banks of a river or lake (Voormolen, 2008). *Homo heidelbergensis* lived from around 500,000 years ago and this site dates from around 350,000 years ago, although the dating has been contradictory and this date isn't conclusive (Voormolen, 2008). How could a *Hominin* allegedly without symbolic culture and by extension without language, have been able to communicate and plan in order to kill numerous individuals of a large dangerous species? The hunting tools found at the site were wooden spears (Voormolen, 2008). They weren't even tipped with stone. This is one of the many mysteries of the Palaeolithic. How did these people live? Did they truly lack representational and abstract communication/expression? Did they have another form of communication/expression that we are not aware of yet? There is the chance that they were symbolically active, but just didn't leave any evidence behind. The problem with this is that as soon as modern humans appear, symbolic culture such as etchings, carvings, cave paintings appear all over the place, and there's relatively quite a lot of symbolic material culture that we have found compared to for previous groups.

Perhaps social organisation and levels of population density had something to do with the high amount of potentially symbolic artefacts being found in the late Middle Palaeolithic and late Middle Stone Age. When there's high population density a larger interconnected social network emerges, this can lead to less chance of inbreeding depression, greater expression of identity, more people to trade and give gifts to, more people to rely on in times of need and can eventually lead to the emergence of hierarchy and inequality (Moreau, 2020). However, the evidence shows that at the time of *heidelbergensis* the population density was relatively quite low, we only see high hunter-gatherer population densities with homo sapiens hundreds of thousands of years later (Wengrow, 2015).

The period after *Homo heidelbergensis* is also very interesting, since this is where the split from *Heidelbergensis* occurs and early Neanderthals and early *homo sapiens* appear (Stringer, 2012). These can be classed as species or subspecies depending on who you ask. Homo heidelbergensis is the last common ancestor between *Homo sapiens neanderthalensis* and early *Homo sapiens sapiens* (Stringer, 2012). I class Neanderthals and early modern homo sapiens as the same species but separate subspecies because, for one, they could produce fertile offspring (Han, 2016), which is the biological definition of a species (according to some) (Shapiro *et al*, 2016) and two, I think that behaviourally they were different, but not that different. Another reason why I find this period interesting is that symbolic material culture starts to emerge, including cave paintings and mobile art (Coolidge *et al*, 2009). This along with more sophisticated burials, and the same consistency of evidence of care for the sick and injured as we see in the deeper Palaeolithic (so care-giving apparently goes further back in time than symbolism), make it a fascinating time (Spikins, 2021).

How did these people think? What was it that drove the move to increased symbolism? Are we missing evidence? These issues will be investigated in the coming chapters. A key question that is explored here is did the emergence of symbolic culture (and by extension language) occur through revolution or gradual change?

The way the species split occurred is that a population of *Heidelbegensis* became isolated in Europe and a population of *Heidelbergensis* became isolated in Africa. There was no (or not much) gene flow between the two groups and different evolutionary pressures caused both populations to develop differently. The northern population was more adapted to the cold so became more stocky and robust which conserved heat (Steegman *et al*, 2002), while the southern population retained relatively gracile characteristics (Ruff, 2002), was taller, and most importantly, eventually developed complex technologies such as tailored clothes (Mcbrearty, 2002), and other innovations such as harpoons (Osipowicz, 2019) which shows they were fishing, and expanding their subsistence strategies. This would have given them a greater chance of survival in a constantly changing world.

It's debatable whether Neanderthal's noses, which were large, were adapted for the cold, but one way of interpreting it was that it was an adaptation to heat cold air as they breathed in (Steegman *et al*, 2002). It could be argued Neanderthals maintained smaller social networks since their population density was low (Bocquet-Appel *et al* 2013), whereas in Africa people may have lived at higher population densities and have been more interconnected (Moreau, 2020). This higher population density may have changed how selection pressures influenced these peoples and culture may have emerged as a buffer to natural selection. In other words, social selection meant that the environment and fauna/flora that existed in it weren't the only influences on our evolution. Human culture affecting how natural selection put pressure on survival in populations may have caused symbolic culture to blossom and spread, together with the technologies, innovations, and belief/ communication systems that came with it (Dean *et al*, 2013).

The African group eventually became successful enough to spread geographically, from around 80,000 years ago, and reached Europe to come face to face with our Neanderthal cousins around 40,000 years ago (Conard, 2006). However, at some point around this time, a process occurred which made most humans living at this time more fragile physically, heavily reduced the brow bridges of both males and females, and may also have given us better skills in dealing with interpersonal matters and issues of debate and discussion. This process is known as self-domestication (Gleeson, 2019).

Anatomically Modern Humans and Gracilisation

We can see domestication in many species that humans use such as goats, sheep, cows, pigs, and dogs (there's a debate about whether cats are fully domesticated) (Serpell, 2016). This process usually results in certain predictable traits (Gleeson, 2019), such as; gracilisation (a less robust form compared to the wild counterpart), smaller teeth, and pronounced variation in coat colour (Gleeson, 2019). It also effects behaviour, with domesticated individuals being more 'tame' and easy to socialize with people (particularly with dogs). With cats it is thought that, unlike dogs, they were selectively bred only relatively recently, and for most of their association with humans they lived largely independently from us, but found a successful niche in living around areas where grain was stored and eating the mice and rats that fed off the grain (Serpell, 2016), Because of this cats may not have gone through as strong a selective process for higher social tolerance towards humans unlike dogs.

Due to archaeological finds, theories have appeared in relation to the domestication process and the evolution of our species *Homo sapiens sapiens*. The 'self-domestication' theory (Gleeson, 2019) puts forward the idea that at a certain point in human evolution, big, aggressive males started being less reproductively successful compared to other group members of the same sex who may have been smaller, perhaps more intelligent, or funny, or better hunters etc (basically some other more subtle attractive qualities began to be selected for). Over many generations, this resulted in the human form becoming less robust, less thick boned, and less muscular. As with many domesticated mammals, it is thought this process also made people 'friendlier'. If we take a dog as an example, it is much more friendly and pro-social compared to its wild wolf counterpart. This theory makes us think, what were the robust, wild forms of humans really like? Were they really knuckle dragging 'cavemen'? Did they behave like us or were they vastly different? The process of self-domestication may be linked to sexual selection. Generally, in the animal kingdom, because females invest more in offspring, it is accepted that in terms of attracting a mate, females choose and males compete (Kuijper *et al*, 2012). In humans, when the process of self-domestication began it may be that females formed coalitions and decided to reject big, aggressive males and instead partner up with smaller, funnier, or more intelligent and articulate males.

This female coalition idea is borrowed from Knight and Power's 'Sex Strike Theory' (Watts and Knight, 1992). This theory argues that at some point in recent human evolution females formed coalitions and denied sex to males unless they brought them meat. This made it less likely that a lone, aggressive male could dominate females and it is implied that males then had to impress females in other ways, and art and music may be a result of males having to impress females in new ways. The process of self-domestication definitely had physical effects on the form that extant humans have today (mainly making us less robust) (Bednarik, 2008) but also may be the reason we are so rational and can discuss and debate calmly. Whether the 'symbolic revolution' occurred before or after is still an open question, and self-domestication was a long process, influencing different populations at different times. In my opinion though, I believe even before the process started, we were a symbolic species, since evidence of symbolic culture stretches back to the Middle Palaeolithic (Burdukiewicz, 2014), to late Neanderthals, who were one of the most robust forms of humans there has been.

Even once fully anatomically modern humans emerged, there were still a few other 'species' of archaic (ancient) humans in a few isolated pockets around the globe. This includes an especially interesting one which flipped the whole of the general understanding of the path of the evolution of anatomically modern humans and even *Homo erectus* on its head. The species which I'm talking about is *Homo floresiensis*, a small 3 ft high hominin with a tiny brain yet association with relatively complex stone (lithic) tools (Groves, 2007). It confused scientists because if this was indeed a branch of the *Homo* line and wasn't a diseased modern human, then how could a population with such archaic anatomy and small brain be associated with relatively recent tools? Does this imply they had language? There are myths from the island of Flores of which the local people talk of little men and women who were small, strange and used to steal and eat babies (Forth, 2008). Was the origin of the myth from the living population of these controversial fossil finds? There is much more that can be said about *Homo floresiensis*, for example that it lived until as recently as 12,000 years ago (Larson, 2008), a time when humans in other parts of the world were starting to grow crops (Tan *et al*, 2022). If so, then this archaic human group could have come into contact with modern *Homo sapiens* who were also living on the island.

The process in which *Homo floresiensis* became so small is known as island dwarfism (insular dwarfism) (Hayashi *et al*, 2020). It occurs with many species, including the pygmy elephants that *floresiensis* used to hunt (pygmy elephant bones were found very near where the hominin bones were found) on the island of Flores (Mazza, 2007). On small islands there sometimes isn't enough biomass in the environment for a creature to maintain it's large size, so over time, smaller size is selected for as individuals with large bodies die off as they can't sustain themselves on a limited diet. This may be what occurred with *Homo floresiensis*. There's also a phenomenon called island gigantism (Van de Geer, 2021), where small species become bigger than their mainland forms. This occurred with another one of *floresiensis's* possible prey, a species of giant rat (Zijlstra *et al*, 2008). What happens is that the small creature becomes isolated on the island, and as there are ecological niches free higher up in the food chain, and abundant food available (relative to the animal's small size), the creature gets bigger and fills the free ecological niche. Another example of this is the extinct giant hedgehogs of some Mediterranean islands (Villier *et al*, 2013).

If *Homo floresiensis* had symbolism it makes us ask yet more questions about humanity and the origins of language/abstract thinking. Despite the reduction in cranial capacity, *floresiensis* seems to have behaved in an advanced, complex way, and used sophisticated tools, despite it's brain not being much bigger than a chimpanzee's (Brumm, 2006). What this points to, is that it's cranial organisation that counts, not cranial capacity as has been enshrined in the dogma of evolution for a long, long time (Holloway, 2009).

Moving on, there are a few other species/subspecies worth mentioning that may have existed at the same time as anatomically modern humans. For example, Neanderthals, from Eurasia physically co-existed with anatomically modern humans for at least 20,000 years in the Middle East and possibly other parts of Eurasia (Wall *et al*, 2013). Separated from AMH by a more robust form, heavy brow ridges, and possibly less of a capacity for symbolic expression. *Homo neanderthalensis* or *Homo sapiens neanderthalensis* (if we use the subspecies definition) is slowly losing its brutish stereotype as archaeologists and evolutionary anthropologists find more and more evidence of the sophistication of the group, especially in terms of hunting/subsistence strategies, and also in terms of symbolic/artistic expression (Peeters *et al*, 2020).

Two other species, of whom anatomically modern humans may have lived alongside, or at least at the same time as are the Denisovans (*Homo denisova*, or *Homo sapiens denisova*) of which we only know about from a fragment of finger bone, but who's existence is predicted by genetic reconstructive maps of human populations which incorporate possible extinct archaic human groups into the picture (they may have persisted until as late as 14,500 years ago) (Derevianko *et al*, 2020). Also, *Homo longi* (*Homo sapiens longi*) who's very well-preserved skull was found to date from as recently as 146,000 years ago (Ji *et al*, 2021).

This date puts it quite a bit before the conclusive emergence of behaviourally modern humans (as symbolism allegedly hadn't fully emerged yet), but after the emergence of anatomically modern humans. I wanted to include *Homo Longi* (known as the 'dragon man') (Ji *et al*, 2021) to show that there were a great diversity of archaic humans from an era not too different from our own. Also because of arguments over where the line should be drawn over naming new species, subspecies, and important populations as *Homo Longi* is a recent new find and addition to the human story (that may eventually be seen as just a slightly different *Homo erectus* or *Homo heidelbergensis*). While many different fossils are found, categorizing them taxonomically can prove difficult, especially when taking into account morphological variation within a species.

From my understanding, the key timeframe, where behaviour changed from 'archaic' to 'modern' is the Middle/Upper Palaeolithic (middle/late stone age) transition, around 45,000 years ago (Tostevin, 2000). This is when we see consistent expression of art in the archaeological record which is, among evolutionary anthropologists, thought to be associated with 'abstract' thought and use of metaphors which could be considered key to how humans act and behave (Galetova, 2019). This book is going to try to show that really, humans today, are using metaphors and symbols that have no basis in reality in a similar way to how we think some people are 'crazy' for saying things that we don't understand. We are apes that have gone mad. Maybe there is no such things as a crazy person. We are all crazy, since we all believe in things which don't physically exist. In other words, we are apes that think in metaphors and use language to create meaning out of symbols that have no real meaning. This is a paradox. We can understand each other by signalling things which are an abstraction and have no concrete basis in reality and only have meaning because we've decided they do (part of the definition of the symbol). We are mentally divergent monkeys. The next chapter focuses on compassion in non-human mammals and how the seeds were first planted for our highly social, yet highly mad way of life.

CHAPTER 2: BEHAVIOURAL SIMILARITIES BETWEEN HUMANS AND SOCIAL MAMMALS

At an early stage when learning Darwinian evolutionary theory, students get taught Zahavi's costly signalling theory (Zahavi, 1997). This is that, signals, whether true or false, have to be highly costly to express, otherwise they wouldn't survive in nature (Zahavi, 1997). This concept is difficult to understand and difficult to explain. However, a good example is that of a 'stotting' gazelle. Stotting is an exaggerated jump that certain deer and antelope species perform when avoiding danger/predators. The animal jumps higher than necessary to say to any potential predators 'look how high I can jump, I can afford to expend this extra energy in order to get away from you, so you better not waste energy trying to catch me'. This behaviour would have been selected for over generations because it works in putting predators off a chase. However, it is only evolutionarily stable because it is costly. If the signal was false, and the antelope couldn't afford to expend the extra energy, then it would all be for nothing and the gazelle be food for the lion. This can exist with many kinds of traits. For example, a peacock's extravagant tail feathers. The male peacock is saying to females 'I'm so genetically fit that even with this huge handicap of a massive tail, I still survive'. Again, this works and is evolutionarily stable because if it wasn't we wouldn't have male peacocks with such extravagant tales. Nature would have selected for a reduction in tail feather size.

In theory, the same is true with any type of signal, including using symbols to represent things in order to communicate (Shaver, 2016). This presents an evolutionary problem, because language is cheap, in the context of the previous paragraph, it doesn't require much investment to lie for example (the equivalent of the gazelle failing to stot and being eaten). So why does it survive and how did it develop?

Language as an extravagant trait due to sexual selection pressures seems an unlikely solution to the problem of costly signalling, and language emergence. There is no difference in language ability between the sexes, which you would expect to see if the 'choosier' sex had had a strong preference for elaborate or otherwise attractive language skills. One aspect of language though, the ability it gives us to more easily form close bonds and friendships, can be seen in many other social mammals including; lions, hyenas, wolves, elephants, and non-human primates. Obviously, these mammals don't have complex language but can still be devoted to each other and show compassion, so language isn't a prerequisite for highly cooperative bonds to emerge in social mammals (Smith, 2016).

Elephants are particularly interesting in respect to compassion and human-like behaviour because they seem to even mourn their dead (Douglas-Hamilton *et al*, 2006). When a group of elephants come across another elephant's skeleton they gather in a circle and just stay there for a while (as if mourning) (Douglas-Hamilton *et al*, 2006). This may imply that they have an understanding of the concept of being dead, and, this is a stretch, but in my opinion, due to this behaviour, elephants may even have ideas related to a potential afterlife. They also, like archaic humans, are keen care-givers (Bates *et al*, 2008). They will often help to the best of their ability any fellow elephant stuck in mud or sand, and often are successful in freeing them (Bates *et al*, 2008). elephant groups are matriarchal, this means they are lead by females (Mcomb *et al*, 2001). There is a trend of many highly compassionate social mammal species being matriarchal. African wild dogs, bonobos, orcas; all are female led and are highly compassionate towards those in their group (Borrego, 2016).

Orcas (Killer Whales) are interesting for another reason in the context of this book. One of only a few species in the mammalian taxonomic category to live significantly past menopause (along with humans) (Whitehead, 2015), it is thought that this phenomenon is a product of an evolutionary process called inclusive fitness. Inclusive fitness is the idea that, in a group, the individual shares genes with his/her kin (Trivers, 1974). So, depending on how related one is to another, if a relation helps the individual to survive and reproduce, then the relation's genes will also be passed on (those genes that the individual shares with the relation), so even if that relation never reproduces and dies young, the evolutionarily beneficial characteristics of that individual will be passed on in the related individual's progeny. This is also used as an argument for why homosexuality is common in humans (because these individuals were beneficial to their group in our evolution and helped relatives survive and reproduce). There is another, similar theory for why in some species females live past menopause. It is known as the 'grandmother hypothesis' (Hawkes & Coxworth, 2013). The argument is that, in our evolution, older women knew where starchy tubers could be located and how to excavate them, provided support for child-bearing female groups members (known as alloparental care), and could give valuable advice and knowledge to conspecifics (group members), among other evolutionarily adaptive skills (Hawkes & Coxworth, 2013).

Because of the previously mentioned inclusive fitness idea, older women were extremely valuable to human group survival, so the phenomenon of living past menopause emerged. Similarly, the same might be true in orcas. They are known to be highly intelligent and their songs may have similarities to human language (Filatova *et al*, 2015). However, if human symbolism is so important and valuable to modern humans (and maybe other members of the *Homo* genus), then why do we not see similar expressive systems in chimpanzees or bonobos?

Chimpanzees can learn to sign (human sign language) (Jensvold, 2014) and can even sign some symbolically representational words such as swearing at someone (through sign) when something upsets them (Jensvold, 2014). However, in the wild there is no sign of this kind of expression (Cissewski *et al*, 2021). It's like the great apes only behave with symbolic intention when they are around humans, which implies they do have the capacity for more consistent symbolic communication (ie language), maybe even among themselves (rather than when interacting with humans). The problem is that in the wild there isn't a reason for them to behave like this, their societies and interactions are completely different from that of the human world so even if they have the capability, another ape wouldn't reply in the same way. Even in humans, language is a bit of an evolutionary mystery.

Those that study the emergence of human culture often regard human language as a part of symbolic culture (Theis, 2010).

However, if language was so closely tied to the symbolic explosion of the Upper Palaeolithic, around 40,000 years ago, then how could archaic humans, from 200,000 years ago, (almost ten fold further into the past) have planned and communicated in order to kill numerous wild, dangerous horses with only wooden spears (Voormolen, 2008)? I refer to the Schoningen (Voormolen, 2008) site (again) in Germany where there's compelling evidence that *Homo heidelbergensis* trapped a group of Palaeolithic horses by the banks of a river and dispatched them using only wooden spears that had no lithic (stone) points attached to them. There's a puncture mark that is consistent with a wooden spear on a equine scapula (shoulder blade) that was found (Conard *et al*, 2015). Along with a portion of an actual spear that was also found (Conard *et al*, 2020). Presumably it would have taken great planning and cooperation to succeed in such a dangerous hunt.

Homo heidelbergensis is widely thought to be the common ancestor of both modern humans and Neanderthals, but from the time that these people were alive there is scant evidence of potentially symbolic artefacts. However, that doesn't mean it's not possible that these populations were symbolic since there are some findings from this period (and before) (Coolidge &Wynn, 2009) which may indicate some symbolic capability. But maybe, *Homo heidelbergensis* and others got by on more simple communication systems such as gesturing, or that they used a precursor to language.

The hyoid bone, when it's found at sites containing archaic human remains is often used to judge whether a population was capable of speech (Capasso, 2008). This is logical, but symbolism could be expressed in other ways. In other words, a highly developed hyoid bone isn't necessary for a species to be able to express themselves symbolically. A good example of this is sign language which is just as symbolic as spoken language. So ancient forms of humans could have rationalised and communicated symbolically without speech. However, the question is, would this have been likely? If we turn, again to *Homo Floresiensis*, we can see that perhaps underdeveloped anatomy shouldn't be used to judge symbolic or technical capability, since they have traits associated with both *Australopithecus* and *Homo erectus* yet may have used stone tools (Argue *et al*, 2006). However, many social mammals are highly caring, compassionate, and cooperative without symbolism, so even though symbolism is very much exclusive to humans that doesn't necessarily mean it could be the only form of 'higher' expression of communication and expression that could develop.

In regards to the social carnivores, all have interesting social systems with African wild dogs and spotted hyenas being particularly interesting. Both display alloparental care (Creel *et al* 1997), (Knight *et al*, 2008), the behaviour of group members other than the parents helping to raise young. African wild dogs are also highly caring and will regurgitate food for ill or vulnerable group members (Breuer, 2002). This idea that they are highly caring/ cooperative is interesting because they also have a high hunting success rate compared to other mammalian carnivores (Fuller, 1993). This implies that there is a link between strong bonds or high levels of cooperation, and high levels of hunting success. Could early humans have been similar in terms of hunting success rate because of their high levels of cooperation?

Spotted hyenas reproduce in an interesting way. Females have what can be described as a pseudo penis, and females must be fully willing to mate when reproduction takes place (Cunha *et al*, 2003). Furthermore, females are dominant over males. Also, Hyena society is matriarchal, status is passed down through the female line (Holecamp, 2019). This is not uncommon in social carnivores (Holecamp, 2019), and its presence shows the great diversity in social systems among social carnivores, which potentially could be a reflection of how archaic human groups operated in the past. Even in modern human societies and populations, there is great diversity in social organisation. The nuclear family is very much not the 'natural' way humans have lived in the past, and was really introduced with the industrial revolution. The support network extended families give is much more akin to living a hunter gatherer lifestyle than the nuclear family and is hence more 'natural'.

Humans have evolved to socialise with 10 or 20 close friends and family for the majority of the time. Ethnographic studies reflect this (Hill *et al*, 2011). In a hunter-gatherer environment we would have had our group to rely on psychologically and socially. Teasing and humor would make sure dominant individuals didn't bully and oppress others (Gray, 2009). This would stop someone's ego building too much. There was always an open ear, or open arms for a hug. This support network would be much better for one's mental health than the nuclear family or even an extended family. A lone individual can be so vulnerable in this society and the institutional state apparatus of society are more likely to brainwash lone individuals since they have less minds available to help and rely on each other. The government and media want to brainwash us into thinking that a nuclear family is natural when in reality it is the optimum unit for brainwashing and oppression.

Karl Marx and Friedrich Engels argued the nuclear family was the start of women's oppression and basically clasped people in chains because that social unit is easy to manipulate and control through the media and institutional state apparatus (Brown, 2013). Similarly to social carnivores, modern humans need social support to thrive, need comfort at times, and suffer when being exploited or ostracised. The difference is that wolves, hyenas, or African wild dogs don't have implicit, covert systems in place in which to ensnare vulnerable or rebellious individuals and maintain the status quo. Humans are so hyper social that these implicit systems may be needed for society to function, despite how unjust it may be. These societal mechanisms must have emerged as civilization developed (slowly over time) but whether these mechanisms had/have been purposefully directed is still rather mysterious and unknown.

CHAPTER 3: EVOLUTION OF SOCIAL ORGANIZATION:

Since the dawn of civilization leaders and institutions have come across uprisings, rebellions, and conflict due to the number of individuals living in one place, the spread of power among the people, and the oppression of certain groups (Armit *et al*, 2006). I say 'come across' because I don't think that leaders, or the unconscious entity of an institution, want or know that societal populations can or will rebel. But with all the variables in a community of diverse beliefs and behavioural inclinations rebellion due to oppression is likely to occur eventually. *Homo sapiens* are hyper social (Richerson, 1999), as we can see any morning on a tube or metro station in many parts of the world. Other apes, especially our closest relative, chimpanzees would not be able to tolerate being together in an enclosed place, and would in all likelihood, rip each other to shreds if put in those situations. Even our very recent ancestors living in pristine natural environments would have difficulty coping with all the people in our major cities (because cities are a very recent development, and we ourselves may not even be fully adapted to them yet). Perhaps this is where anxiety, depression, and psychosis are rooted. The psychological stress of living in such great numbers, and interacting with so many different people, not always in a positive way, might be damaging to people's 'mental health'.

However, modern humans have developed both biological and cultural adaptations to deal with interacting with a great number of people. Language means we do not have to act like 'animals' and don't have to live by a dominance hierarchy, and kill each other, but can reason and debate when there is a problem. Culture and a sense of belonging and love due to the community may act as a buffer for the oppressive forces in the struggle of human life.

But let's go back a bit. If we look at the Mesopotamian early civilisations, this is allegedly the first-time people moved from nomadic lifestyles into permanent homes, exploiting domesticated animals, growing crops, producing art and music, and having powerful institutions which held power (Pouyan, 2016). This is much different to, hunter-gatherer tribes moving around the landscape, hunting game, fishing, and exploiting fruit and vegetable resources. These communities were interconnected (from at least as far back as the Upper Palaeolithic) and were less isolated than in the times of the Neanderthals, maybe 20,000 years prior to the Upper Palaeolithic (Mellars and French, 2011). Modern humans of the Upper Palaeolithic were 'free' (not controlled or oppressed by institutions). The move to a settled life, and larger settlement populations brought control and oppression by religious and governmental organisations. From this, unrest will inevitably develop and that may be one possible cause of civil war. I'm not saying that war didn't exist before civilization and there is interesting discussion of this (Angelbeck, 2006). Chimpanzees are infamously known to be war-like. In the wild, males will form coalitions and patrol the boundaries of their territory, looking for lone individuals from other groups to ambush (Wilson *et al*, 2004). This aspect of great ape temperament reflects the violence and brutality that humans express during war. Chimpanzees often mutilate or kill any individual, from an opposing group, they find. However, one subspecies of chimpanzee, the bonobo, shows much less aggression.

Bonobos are very different to our equally close relative, the chimpanzee. Their social structure and social behaviour is different. Bonobos are matriarchal, female dominant, and show slightly less sexual size dimorphism than chimpanzees. They are also allegedly less violent than chimpanzees and have sex much more readily (Woods, 2010). They even have sex as a greeting sometimes (Woods, 2010). Interestingly, these two species are equally related to us because their evolutionary lines split from each other after it had split from ours. So which species are we more similar to? The highly sexual, female dominant peaceful ape or the aggressive, stronger, male dominant ape? I think we are somewhere in between and that there is great diversity of human behaviour and structural/hierarchical organisation both in the past and at present. Furthermore, I think the dichotomy of bonobo Vs chimp is a good parallel to left Vs right wing politics. Chimpanzees are the more violent, anti-social ape, and bonobos seem to be more cooperative and less violent. Chimpanzee violence really embodies the potential for violence and atrocity in our line. Gorillas, who are slightly less closely related to us, are also usually peaceful.

The trend of violence in our extended taxonomical family definitely diversified its application with the rise of our genus and the use of tools (Armit *et al*, 2006). There is much evidence of war in prehistory, and the knife, was probably one of the first deadly weapons invented. With the power of symbolism we have innovated with our killing methods. Was this connected to the human psychosis that I'm proposing? Yes, but equally our symbolism and psychosis has spawned great works of ancient engineering, and breakthroughs in philosophy, maths, and science.

Back to early civilization, the inclinations and drives for society to be subjectively ideal must have caused friction and conflict even back then. As civilization became more and more complex, with a greater diversity of cultural groups and religious sects. Institutional state apparatus may have caused inequalities to develop, and mechanisms may have (naturally?) appeared which curtailed universal human rights and basically sacrificed people. This may have been deemed a necessity to the progress of civilisation. However, are there spiritual undertones to this. What if sacrifice, chastity, blessings actually do influence the real world?

Sacrifice would obviously influence the real world in terms of the person or animal being slaughtered losing their life. What if you traced the line of influence. To the person's mourning family, to the baker who was late for work because he was depressed at his son's death, to the painter who went hungry because he couldn't buy his bread, and so on, and so on. Then, wouldn't it show that spirituality does actually exist? Could it be possible that the inequality and injustice may proportionally increase as civilization and wealth increases in a society/civilisation. Maybe life in the past wasn't as 'red in tooth and claw' as some of us currently think (Weiss, 2010) and life is actually harsher now?

Witches being burnt at the stake in the Middle Ages and literal human ritual sacrifice in South America come to mind when we think of human sacrifices. The media portrays modern western society as different to the violence and barbarity of the middle-ages but persecution, oppression and violence still goes on (but behind closed doors). The ancient Greeks and ancient Romans were open about the violence, traps, struggles, and strives of their era, and they achieved such amazing advancements in science, philosophy, maths, engineering and more. They knew how to craft a person from birth and I feel this has been lost in modern times in favour of lies, exploitation, contradiction, and indifference.

Whether people were happier in the distant past compared to now is debatable. Obviously, since we can't go back and ask them we have to use surviving documents and archaeological evidence to build a picture. In modern times covert traps appear which chastise and oppress people, mostly unjustly, for rebelling in one way or another. These systems ensnare individuals and put shadow on hope. The power of labels can limit someone's life chances, especially in the western world (Chapter 8). I'm always wary of condemning anyone, because I don't feel I have a right to, neither should others.

There is stigma now about spiritual expression in the Western world. However. if we look at the concept of karma, which I define as an energy flow in which morally good actions are rewarded and morally bad actions are punished in an endless flow, then it would be possible for institutions, groups, or individuals in society to build up energy (positive or negative) through the labelling, and oppression of groups or individuals. This could be used (perhaps not consciously) to advance a civilisation into a stronger, more secure position. In other words, using the theory I have just outlined, human sacrifice may actually be 'worth it' and work for the interests of those in power, or in other words, the oppressors. Or, another view, is that I'm just 'psychotic' (like billions of others). To sum up the chapter so far, people have developed morally, spiritually, and symbolically since the dawn of the Upper Palaeolithic at least. We have the potential to be much more social and cooperative than our great apes cousins. The symbolic breakthrough over a hundred, thousand years ago set us on our path, but there have been constant uprisings and rebellions during the birth of civilisation.

Marxist Perspective

Although Karl Marx never wrote grand theories on hunter-gatherers, he did lay out the idea that humans emerged from a tribal stage, to a stage of primitive communism, to a feudal stage, to capitalism (to a hypothetical future communistic phase) (Marx, 1996). Key to understanding Marx is understanding the importance of who controls the means of production (and the actual product and surplus) in relation to subsistence of a population. Marx believed if everyone equally shared the means of production, society would function better, be fairer, and there'd be less injustice and inequality. He thought that the workers in a capitalist system were slaves to the system, and that the factory owners were profiting and getting rich out of others hard work. It was fairer for the workers to own the means of production. This is what communism is based on (although in most of the places it has been implemented it hasn't been purely positive).

The stage of primitive communism in the writing of Marx could be described as similar to Rousseau's view of man in the state of nature (turn to chapter 5 for more on this) (Ryan, 2012), where egalitarianism reigns, and the means of production is held fairly, by all (and in Rousseau's view man is at his happiest). In this case the means of production would be the work in terms of hunting and gathering. People shared everything and there was no concept of private property. There were no tribal kings sitting on top of rare antler jewellery and lion pendants in this view. People hunted, fished, and gathered plants and mushrooms and spent the rest of the time socialising, maybe tailoring and honing skills, and generally having fun. There's no such thing as a utopia though and maybe this view is rose tinted. This social condition of 'primitive communism' would have occurred after the symbolic revolution and humans beginning to think in abstract, metaphorical ways (we can see similar forms of social organisation in some egalitarian hunter-gatherers that exist today).

The development of 'primitive communism' was probably tightly linked to the evolution of modern life histories and division of labour between man and woman and may have co-evolved with the emergence of those two features. Social organisation grows and develops through time, and is subject to social and evolutionary selection pressures. Ideology doesn't have to be consciously involved in the development of a society.

Where one is personally on the political spectrum might be a universal part of human belief in all modern humans, and it makes one wonder what ancient hunter-gatherer views were on the values of the left or right, for example. I guess we could ask modern hunter-gatherers what their views on politics are. This is just something interesting to think about.

The Functionalist (or Neo Functionalist) perspective puts emphasis on how the systems in society work together to create the norms, values, and subcultures of that particular society. Emile Durkheim (an early Functionalist) viewed society as an organism that must stay healthy just like a biological body (Malik *et al*, 2022). In modern society the organs would be the media, education, institutional state apparatus (such as the media), repressive state apparatus (such as police, army etc). In hunter-gatherer societies that could be the division of labour, education (mainly informal), and hierarchy (if there is hierarchy in that society). However, as you might be able to see, the Functionalist perspective breaks down a bit when looking at hunter-gatherer society because tribal societies are fluid, and there are few 'institutions' if any. There is usually no centralised power. What could we take from this? Well, as hunter-gatherer societies are highly flexible in their social organisation, they do not need grand departments and institutions to keep control because there are a lot fewer people. In hunter-gatherer society the cultural norms, values, and rituals are usually enough for people to be controlled and content. I don't mean controlled as in oppressed, I mean in terms of the people being happy members of the group, with no reason to rebel or reject values (unlike in Western industrial society). I'm even apprehensive to use the term individual when talking about hunter-gatherers. Individuality almost doesn't exist in egalitarian hunter-gatherers which is a reflection of the 'primitive communism' label that Marx used to describe hunter-gatherers.

According to Marxist philosophy once hoarding of the surplus (goods, valuables, tools) begins to occur, inequality will emerge (O'Hara, 2008). Unfortunately, and ironically, bursts of symbolism (for example art and music) actually increase when inequality emerges due to surplus being hoarded (Wengrow and Graeber, 2015). In a nutshell, the theory of how this occurs is that egalitarian hunter-gatherers start to become very successful in acquiring valuables (food, animal hides, tools, art), then the surplus begins to be hoarded, and certain individuals profit from it more than others. This leads to a system of hierarchy. There are those that are materially wealthy and those that are not. The rich individuals then have others create art, music, and other material culture to express their power. However, I don't mean to be so negative about the world. There are many interesting optimistic things about hunter-gatherer social organisation.

This includes that children are physically held and picked up more in hunter-gatherer communities compared to urbanized environments (Narvaez *et al*, 2014)). Plus, hunter-gatherers are reportedly less likely to become depressed than those that have grown up in cities (Frackowiak *et al*, 2020).

Hunter-gatherer social organisation can be flexible, and groups can change from more egalitarian, to more hierarchical based on the season. For example, in the arctic, some hunter-gatherer peoples change from fully egalitarian and nomadic during the summer (hunting Reindeer), to sedentary and hierarchical during the winter (when food from the sea is more heavily relied on) (Kelly, 2007). This shows that we shouldn't label human groups as of one kind or another, as human groups are fluid and flexible, and social structure can change often. From a Marxist perspective I think this fact adheres more to a Marxist perspective than it does to a Functionalist one as it shows the fluidity of hunter-gatherer life and how people could change social organisation based on optimization of control of means of production (or in other words trying to maximise their hunting/gathering rewards). There are few functional, powerful individuals or groups involved in how the society structures itself. Hunter-gatherer life is much less structured than modern life.

What can we take from this discussion? Well, maybe we should not be living such regimented, strict lives, where we work in an office most hours of the day. Society should allow us the options to relax, and follow what pursuits we want to follow, as in a hunter-gatherer mind set. We are not adapted to be slaves to capitalism. Furthermore, the more surplus that is hoarded because of greed in this society, the more likely deeper inequality is going to emerge (if we use inequality that emerges due to surplus hoarding in hunter-gatherers as an analogue)..

A Rough Timeline

For the sake of simplicity let's breakdown some eras in the evolution of human social organization. The era of early *Homo erectus/heidebergensis/neanderthalensis* 200,000-70,000 years ago. Then the era where a diversification of subsistence strategies and technologies emerged 70,000-20,000 years ago. Then the beginnings of agriculture, and explosion of art 20,000-8,000 years ago. I find the era between 20,000-8,000 years ago very interesting because the material evidence we have from this era shows the great human potential for innovation. This was the African late Stone Age and the Eurasian Upper Palaeolithic. Faunal assemblages at human sites from this time show a greater diversification of types of animals, there is more sea food taken (Verdun, 2021), and technology such as the harpoon (Estevez, 2013) and the needle appear (d'Errico *et al*, 2018). Most groups of people had been through the self-domestication process and social life was quite similar to hunter-gatherers living today (possibly). We'll come back to this exploration of early human life, but now let's move into how intelligent species from other taxa could show similarities to humans. It might make the point that humans are mad, more obvious.

CHAPTER 4: CORVIDS AND ORCAS LESSONS FROM DISTANT RELATIVES

Convergent evolution is the idea that organisms can converge in form, behaviour, or biology due to similar selection pressures making them evolve along similar paths. They do not have to be closely related and a good example of this is the *thylacine*, or Tasmanian tiger, which is a recently extinct marsupial, that evolved a very similar form to the northern wolf (*Canis lupus*) because it filled a similar ecological niche (Figueirido and Christine, 2011). It could be argued, the same can be seen in highly intelligent but different taxonomic groups of animals in relation to tool use and problem solving. *Corvids* (crows) and orcas (Killer Whales) are noticeable contemporary groups that have converged with humans in this respect (Rolando, 1992), (Hill *et al*, 2022).

Orcas have similar life histories to humans, with females experiencing menopause. This suggests, that they are highly social and that selection pressures have influenced the survival of females past the point when they can have offspring. If this has occurred in a similar way to humans, it is because the females have valuable knowledge and can help their own offspring raise their young (the post menopausal females' grandchildren), (also known as alloparental care). Furthermore, Orcas have complex communication systems and dialects that pass on from generation to generation (Rendell, 2001). This is a form of culture. However, it brings up the question of symbolism. Do orcas communicate symbolically? What actually is symbolsm (turn to chapters 6 and 7)? I would define culture as 'the passing on of social or environmental information from generation to generation through a learning process that is independent of genetic transmission (yet can eventually influence the selective pressures that create genotype)'. There is no universally agreed upon definition of culture. It doesn't have to include abstract or arbitrary signals which are agreed upon by the whole group (which is a definition of symbolism). In other words, the culture expressed by cetaceans is probably not symbolic (although it could be).

Orca dialects vary based on geography (Rendell, 2001). Dialects of their song are different based on what you part of the world they inhabit. Ethnographic accounts seem to be more useful than experimental approaches in the study of cetacean culture (Rendell, 2001). Male humpback whale song seems to be uniform over vast distances of the ocean basin and vary mainly based on time rather than space (Rendell, 2001). There is some suggestion that the uniform song begins when male conspecifics copy the song of a near by dominant male, although there isn't overwhelming evidence of this (Rendell, 2001). Again, there seems to be no sign of the song being arbitrary, which means having no similarity to what it represents, and coding for agreed upon concepts or things. The emergence of symbolic culture, rather than simple culture may be the point that animals reach our level of crazyness. Humans can use arbitrary signals (speech) to each other at a rapid rate, sparking off imaginations and mad thought processes that we so far think that no animal is capable of. However, we could be wrong.

Corvids, in the UK, include the very common carrion crow, which can live up to 30 years (Baglione *et al*, 2016). This long lifespan, similarly to modern humans, means individuals have a great amount of time to develop skills and experience of their environment and observe the behaviour of others (Baglione *et al*, 2016). It is said that carrion crows can identify individual human faces that they've seen before (Brecht *et al*, 2017), which is remarkable not only for the length of memory a crow has, but for a species in a completely different taxonomic *order* to perceive this. I've often seen crows in my neighbourhood and thought they were watching me, or at least knew of my presence. While this may be paranoia on my part, there is no doubt that they are highly intelligent, including being able to solve puzzles that some apes and even small children find difficult (Heinrich & Bugnyar, 2005).

Crows have a long history of being represented in ancient cultures through art and myth (Jila, 2006). They are often depicted as ill-omens or beings that border this life and the afterlife. They are also often seen as wise or clever. Modern views on the them echo both that they are bad luck, but also that they are intelligent (they can be known as tricksters). *Corvids* are widely distributed throughout the world and seem to have adapted to human civilisation really well, similarly to the red fox.

Marzluff (Marzluff and Agnel, 2005)), argues that during our evolution our close relationship with *Corvids* (primarily at scavenging sites), who can themselves develop culture (the passing on of knowledge from generation to generation), would have impacted and had a feedback effect on the evolution of our symbolic culture. This is termed 'reciprocal adjustment'. Cultures from two distantly related species could have influenced and spurred on the development of more complex cumulative culture in each species. Known as tricksters (Chappell, 2006), the crow's place in our mythology is due to their inquisitive, and mischievous nature.

The association of many species of *Corvids* with carrion could be similar to a stage that our genus went through in the Lower and Middle Palaeolithic. *Homo habilis,* our ancestor, is thought to have scavenged carcasses primarily to get at the nutritious marrow in cracked bones (Dominguez-Rodrigo, 2002). This is often given as an explanation for how meat first started to become a significant proportion of the human diet (slightly before the emergence of cooking). Whether *habilis* did routinely scavenge carcasses for marrow is debateable but by the time *Homo ergaster* appeared meat in all likelihood was a more significant part of the diet as we can see from the more modern form, larger brain size, and presumably shorter gut (Anton, 2003). These changes that we see in *Homo ergaster/erectus* are known as the 'expensive tissue hypothesis' (Navarrete, 2009) which postulates that cooking and eating meat let calories be accessed and processed more easily which resulted in a shortening of the gut (carnivores tend to have shorter guts compared to herbivores) which in turn meant more nutrients were available for encephalization (brain size increase) (Aiello and Wheeler, 1995). The point is that if scavenging carrion was a stage that our lineage went through, then that could be a strong argument for the idea that *Corvid* consciousness, behavior, and intelligence while different from ours may have interesting parallels. In other words there could be some convergence in subsistence strategies and by extension behaviour and cognition.

However, *Corvids* seem to lack symbolism which has been a key part of human life for the last 50,000 years at least (Hovers *et al*, 2003). Symbols have acted as points of common belief for people and has both brought us together and caused conflict. Even the way we communicate (speech and writing) is based on symbols. *Corvids* have various calls, a lot of which we still don't properly understand but they lack a symbolic system of communication

(Roskaft *et al*, 2023). Perhaps if *Corvid* society becomes more complex in the future a symbolic communication system could appear? They have talons which could be used similarly to hands. Or maybe symbolism isn't required for civilisation and a communication system based on something else could appear?

There's other aspects of *hominin* and *Corvid* behaviour that may converge including the existence odd sexual behaviours in both species. *Corvids* are known to copulate with dead conspecifics (Swift *et al*, 2018), a behaviour that is known as necrophilia in the human species. Despite how abhorrent the behaviour is, it is worthy of note in this context as it is another similarity between these two species. I don't know of any cross species comparison on niche sexual behaviours but both in the positive and negative it seems *Corvids* may converge with ourselves. Maybe using human moral values to judge *Corvid* behaviour is anthropomorphizing them to a huge degree. Labelling this *Corvid* behaviour as paraphilia probably isn't accurate, since the cognition and behaviours of *Corvids* are not as well understood or categorized, as in humans. *Corvids* show interest in dead conspecifics, possibly to learn to avoid dangers so that they don't suffer similar fates. They have 'crow funerals' (similarly to elephants), and this may infer that they are thinking about mortality and the concept of death (Swift *et al*, 2018).

Among humanity the western world likes to label individuals based on their sexuality, which can lead to a self-fulfilling prophecy. There is another, more accurate and less harmful way of looking at this issue. That homosexuality and even the paraphilias are something within every human that we don't need to express, but need to manage and help rather than persecute the individuals that express these behaviours. The fact that carrion crows actually express necrophilic behaviour could be linked to well developed skills in other areas, such as intelligence. More research into sexual behaviours of intelligence animal taxa could be really insightful in terms of knowledge of the relationship between sexual behaviour and intelligence. Western society treats the issue of sexuality (and with this psychiatry too) in a rigid, definitive way and labels associated with it are wound up in a complex web of propaganda and lies that corrupt the human condition from our natural hunter-gatherer lifestyle and mind-set.

In terms of other animals with high intelligence from different taxa in the animal kingdom, octopi are probably the animal most different from us with well documented high intelligence. They have been known to solve puzzles to obtain food and can even open tightly closed jars with their tentacles (Richter *et al*, 2016). Their nervous system tissue, in other words brains, are located down their tentacles (Young, 1965) which is really novel compared to chordates (animals with a backbone). octopi don't seem to have complicated social systems. This is interesting to me as human intelligence is often said to be a product of adaptations to our social environment (the 'social brain hypothesis') (Dunbar, 2023). So, if it's not adaptations to interacting with conspecifics which lead to octopi intelligence, what could it be? Perhaps it's interaction with environment considering octopi have such a relatively unusual anatomy?

One theory (Amodio *et al*, 2018), for the evolution of octopi intelligence is that once cephalopods, like octopi, lost their protective shells, the increased predation that they received meant that they had to find novel ecological niches in order to survive. The increased predation, meant that unlike in intelligent vertebrates, long life histories couldn't develop, but the challenging ecological niches that octopi had recently occupied meant that high intelligence was selected for despite short life histories. This is what led to intelligent Octopi.

In vertebrates like humans, dolphins, and corvids, we have had the luxury to develop long life histories (this is especially the case with humans) (Key, 2000). This was probably possible because we protected ourselves from predation well. This means that, for example with humans, we have years and years to learn and also physically grow our brains (during our childhood and teenage years). Cephalopods haven't had this luxury which shows that the cognitive adaptations that they've had to acquire in difficult ecological niches, across short lifespans, is even more impressive.

Elephants, like humans have long life histories (Wu, 2023). Nature has allowed this to develop because elephants are large animals and because of this have protection from predation. This has led to elephants evolving long lifespans. With this they have also evolved high intelligence. As with humans, elephant 'childhood' is a time when they are learning and their brains are growing. This prepares individuals for life as an adult. elephants even pass the 'mirror test' (Brandl, 2018). That is, that they identify themselves in mirrors rather than thinking that what they see is another elephant. Few animals pass this test. This could show a well developed theory of mind, although that may be an oversimplification as the mirror test probably is not a direct analogue for theory of mind. Elephants may know that they are sentient beings and that there are others like them in their environment. The creatures that do pass the 'mirror test' include dolphins, apes, and humans.

The fact that long life history can aid development of intelligence makes one think of how these features developed in humans. Humans do not have massive size to protect them from predation. It may be social cohesion, innovation, and ingenuity which has led to long life histories (and intelligence) developing in humans. We would bond, plan (with language), and build defences against predators. This planning is such an amazing unique feature in the animal kingdom. No animal has evolved anything like human language. However, I believe the breakthrough into the symbolic realm could be described as a 'psychotic break' because the symbols we use have no solid basis in reality. It all exists only because of group agreement of what symbols mean.

It's often argued modern humans are the most intelligent, sentient, and advanced of all the creatures in the animal kingdom. This is put forward by religious institutions, general societal norms and values, and the media. However, new discoveries showing we are not unique in our special traits are constantly being made. For example, we now know that tool making is found in various taxa, whereas once, it was thought to be unique to humans (Bentley-Condit and Smith, 2010). This shows that we shouldn't take it for granted that we have a more sentient reality than other animals. We are hypocritical in the modern condemnation of hunting, yet we raise chickens, pigs, and cows in tiny enclosures and poor conditions just to be slaughtered. Would it not be better to let an animal live a free life and have a chance to escape before we kill it to eat it? The intelligence of other animals and the sophistication of their communication systems, even when they are from diverse taxa, shows the power that evolutionary pressures can have in terms of potential for convergence. Western society uses systems and lies to trick the young minds of it's society into believing the dogmas that it projects. Humans have lived a hunter-gatherer lifestyle for 95% of our existence. Were people happier then? We don't know. However, much of what is said in the media and taught in schools today are specific to the society that we live in. The next chapter will look more closely at Western society and how 'man in the state of nature' is viewed.

CHAPTER 5: WESTERN SOCIETY AND 'MAN IN THE STATE OF NATURE'

The 18th century philosopher Jean-Jacques Rousseau was an 'age of enlightenment' thinker, He had a significant impact on the early modern thinking regarding the nature of humanity.

Rousseau's essay on the origins of inequality (Rousseau 1712-1778, as cited in Hackett 1983) portrayed 'man in the state of nature' as the natural human condition being not that distant from that of an animal. As in, we followed our basic needs of wanting food, water, shelter, and companionship/ a mate. I think that he has left out a key element which separates man from animal, symbolic culture. In our natural state, all of the various forms of social organisation of hunter-gatherers that existed, with our unique symbolic communication system, is evidence that through all the variety of genetic and social diversity we were complex, intelligent, resourceful, social beings. We had a keen need to express ourselves. This was satisfied through the creation of art, music, storytelling and more. Whether the tribe was communistic and shared everything, or more hierarchical with kings and queens, human society had that key element, symbolism, including language.

In the hierarchical, or more complex forms of hunter-gatherers there were even specialists, crafters of tools, weapons, personal ornaments and more. There is no concrete trajectory from simple and communistic to more complex and hierarchical hunter-gatherers but the element of symbolism and religion makes these people masters of their reality with skills useful in the wilderness that would surpass 99% of people that exist today. Religion and spirituality, an element of human life that is still key in many peoples lives, had it's beginnings here, with shamans blessing and healing people in many different ways. The West teaches that religion is wrong, science is the only truth. However, by doing this it is pulling the wool over eyes and hiding spirituality from you in order to better control and manipulate you. If someone doesn't believe in the afterlife and treasures this one life then they are less likely to want to die for what they believe unlike a religious person. Ironically this is the opposite to what Marx said that 'religion is the opium of the people' (Marx, 1843) and numbs the pain of a hard life of physical labour with the promise of heaven after death.

Furthermore, the West likes to use labels definitively. This is in order to categorize people. As many have said, if you divide people then they are more easily conquered. Other cultures across the world have a more open minded way of thinking and know that change is possible. This small minded mind-set of the West has seeped into western psychiatry which is why psychiatric help isn't the answer for many people. A different way of thinking shouldn't qualify as a mental illness. The truth is that the west picks and chooses who has mental illness. We could all be labelled as psychotic because we are all psychotic apes that communicate with symbols that have no concrete representation in the real world.

If we return to Rousseau's arguments, (and as opposed to much of the popular thinking of many of his contemporaries) he argued that 'man in the state of nature' was a happier time for the humans that existed compared to in large cities of the time. People's lives were based on sharing, being fair, and not dominating or subjugating anyone in your community. The tribal lives of many communities 20,000 years ago could be described as 'primitive communism' although there is debate over in which aspects this was true. Hobbes was a contemporary from the time of Rousseau and unlike Rousseau he argued that 'primitive' life was brutal, short, and full of fear. Danger was around every corner and life's struggle was at it's most intense. However, there is evidence that the hunter-gatherer lifestyle does indeed lead to a happier populace (Frackowiak, 2020). Stories from the 1600's of westerners who were raised with hunter gatherer communities but then brought back to western cities and then ran back to their fostered communities are not unheard of (The Independent, 2016). This indicates that the social bonds individuals made through living hunter-gatherer lifestyles meant more to them than living comfortably but in a world where hierarchy and subjugation is the norm. This is evidence that Rousseau's view of 'man in the state of nature' is more socialist and based on equality than Hobbes' theory (not that that necessarily means it's more accurate). The argument that modern, highly urbanized modes of human living goes against our nature and evolution is highlighted by the idea that a hunter-gatherer lifestyle leads to a happier and more mentally stable person.

Leading on from this, Dunbar's number (Dunbar, 1992) is an idea which postulates that there is a maximum number of people that each human individual can effectively socially interact with. This is because for most of human existence we have lived with vastly less people in our social network than is possible now. This is at the core of the west spreading lies about identity. We are not designed for the world we currently live in. Mental illness and rebellious ideas go hand in hand and are a result of this. 'One mans psychotic episode is another mans spiritual experience' if looked at with all evidence that isn't exclusively based on scientific method. By this I mean subjective experience. Anything can be possible with illnesses like psychosis and it brings the whole concept of reality into question. While eastern spiritual thinking would approach these 'conditions' with an open mind, the west would rather stamp a label on a person and give them medication when the solution could be a deep conversation with someone who truly understands. Like the lesson from the original Pokemon movie (Pokemon The First Movie, 1998), crying, and either sympathy or empathy actually help to heal.

Furthermore, western society, and maybe civilisation in general changes the human heart, in that for hundreds of thousands of years we lived based on principles of equality, love, and kindness, and suddenly in the last 2000 years hierarchy, subjugation, oppression, and even torture has been thrust upon us. This is not natural, is not good for who we are or where we are going, and betrays our origins and our very humanity. A servant looking after a household, a factory worker working for a factory owner, a soldier dying for a lieutenant. These are all examples of the submission based on unfair principles which modern society makes us submit to. Not only is this system immoral but it has lead to damaging consequences for the environment including pollution of the atmosphere and destruction of ecosystems due to capitalist/imperialist greed. The system of capitalism which western countries enjoy actually requires the exploitation of poorer third world countries. This is known as economic imperialism and still exists to this day. Figures such as Che Guevara highlighted these issues (Guevara, 2003). His proposal of world communism may have been a good idea, but could it have lead to the replacement of one kind of tyranny for another, as is sometimes the case, as maybe Russian communism was an example of (Brown, 2011). The negatives of communism include banning of religion and less liberty/freedoms in terms of personal rights (if we look at communism as it has existed up to this point).

As a person who has experienced psychosis I have felt flows of energy, what would probably be best described as Karmic energy. My first psychotic episode was a really strange experience. While it was harrowing and scary it also gave me a form of elation and power that I have never felt before or since. I was suffering, possibly due to childhood trauma. I became enlightened to multiple potential truths. Things that you can't be told, but have to learn and understand subjectively yourself. This is why previously I said one man's psychotic episode is another man's spiritual experience. If one continues the logical path of the values of karma it could be argued that karmic principles also apply to society and civilisations. I would argue that on the relatively small-time scale of a human lifetime karma or acts of god can seem like punishments or rewards for ones deeds. This can be amplified by psychosis. This is clearly not scientific, but still has value in terms of subjective experience. I need not go into the historical battle between the dogmas of science and religion. Following karma and generally trying to be a good person are good values whether one believes in God or not.

However, I feel the West is lying to people by mocking religion and it even makes one think of conspiracy theories like the illuminati being real because, to me, it is so clearly obvious that the West's denial of spirituality is a ploy to trick people into submission. Also, the west's alleged acceptance of homosexuality is also a twisted lie. They accept the act, but society still heavily stigmatises people who are openly homosexual. In fact, the level that people label, pigeonhole, and judge openly gay individuals is even more harsh than in the Islamic world. In Islam it is the act that is an offence. There are no systems of persecution and mental torture based on if someone is slightly more effeminate or masculine whether someone is a male or female respectively.

On grand scales, such as civilisations, great injustices and acts of causing pain inevitably occur. I believe that eventually, if we follow the karmic flow of energy principle expressed in the previous paragraph, this will eventually (after many centuries) result in a cataclysm that will damage or even completely destroy that civilisation. It is likely that to stop this occurring based on karmic principles great spiritual sacrifices are made in every civilisation.

Unfortunately, this could be done through isolating and persecuting certain groups, whether that be Jews, black people, homosexuals, sexual deviants, or other minority groups. In the West, the lie is deeper, more hidden, and more extreme than in any civilisation that has come before it. The West tricks people into denying their own homosexuality and then labels, tortures and destroys them once they reach adulthood and realise the truth of human sexuality that we are all on a bisexual spectrum. Furthermore, no help is offered for the anxiety and depression that are a product of this. Even the police don't protect people that are labelled in this way, they turn a blind eye or even encourage the persecution and sometimes physical torture of these unlucky individuals. It 's a conspiracy that runs deep, and there are those that know, and those that don't know.

The countries in the West that are most keen on this form of torture to gain such spiritual power from these individuals and continue to live with wealth, are England, and Spain (that I know of). The UK is the original empire (in a negative sense).

If we go further back in the story of humanity, the beginnings of persecution for sexuality, and the real start of homophobia was at the birth of civilisation in Mesopotamia. Homophobia was laced into religious values and morals. However, the reason for its introduction was not really because it was as disgusting as its proponents argue. It was because homophobia stops revolutionaries from breaking through. This is because, if we lived in a free world our love for both male and female sexual partners would make social dynamics very easy to tip over into rebellion, as love would be felt by many people, and any injustice caused by the state would not be tolerated by individuals or groups, even if that individual desperately tried to control themselves and not express their feelings of love.

Whereas the 'gay man' is submissive, both to other men and society, and will not commit revolutionary acts. There is one theory, put forward by my former lecturer, that human symbolic culture emerged through revolution. Camila Power and Chris Knight developed 'Sex Strike Theory' (Knight, 1991) which postulates that at some point in the last 200,000 years women formed coalitions and denied men sex unless the men hunted and brought them meat, and this solidarity between women in the hunter-gatherer group meant that men had to find new ways to impress their potential mates so began to express themselves via art and music. Thus, symbolic culture began to flourish. This would indicate that symbolic culture first blossomed not because of inequality and hierarchy, but because of sexual selection dynamics.

CHAPTER 6: THE ORIGINS OF SYMBOLIC CULTURE

As has been established, the key difference between humans and other animals is our use of symbolism to communicate and express ourselves. Looking at the emergence of symbolism free from western dogma is not an easy task since those in the west have been raised with this world view. There have been many other world views from past civilisations and the views of the West cannot be objectively viewed as superior. The Ancient Greek philosopher Anaximander had a theory that humans had originated in the sea, from fish (Zuiddam, 2019). This isn't too far from the truth. Although, most ancient beliefs concerning human origins were tales of gods and spirits, more metaphorical than to be taken literally. However, today, with cross-disciplinary studies we can produce some more objective, and scientific theories on how, the essence of humanity, symbolism, developed.

To understand symbolism, as the symbol is integral to language, understanding language is useful. Language allows for an infinite amount of sounds coding for an almost infinite amount of words being produced which allows us to convey meaning and communicate with one another. The evolution of language may have been rapid, or emerged more in stages (Planer and Sterelny,

2021). However, physical evidence of symbolism, in the form of material culture, only emerges in the last 200,000 years (Von Petzinger, 2017). This implies complex language has only been around that long because why would someone speak in metaphor, but not express themselves technologically or artistically? Smith (2009) argues that a genetic change, occurring around 200,000 years ago created the first symbolic human and then this mutation spread throughout the population. I disagree however and feel access to the symbolic realm was a talent that, while to some degree being based on genetics, could have been picked up by any human living in the world at that time. Evidence towards this is that a trained chimp or gorilla can put together symbolic phrases and create their own words and phrases using sign language. A 'psychotic episode' in a member of the genus 'homo' from around 250,000 years ago could have sparked an artistic and abstract trend of expression which, perhaps through luck, conspecifics understood and copied, thus it spread through the population.

However, then one must think, how could humans prior to this have cooperated and hunted, large dangerous game? The Schoningen spear point (Voormolen, 2008) and associated evidence point towards a large hunt of horses being conducted by *Homo heidelbergensis* over 150,000 years ago (as I have previously mentioned). It is thought that these people didn't have language but then how could they have planned well enough to ambush or chase these large, dangerous, wild animals to the edge of a river or lake and proceed to kill and butcher them? Remember a human is comparatively weak and feeble compared to most wild animals, so a horse, who weighs many times more than one individual hominin, and is also many times stronger, and must have been capable of easily killing one of these individuals. Wolves also hunt large dangerous game, but they have sharp teeth and are physically formidable. The hominins were allegedly only armed with wooden spears. I don't think a group of average modern day humans would be able to take down two or three horses if they were only armed with wooden spears, but then I guess they also wouldn't have had a lifetime of training for it.

Just because *heidelbergensis* hasn't, so far been found significant amounts of symbolic artefacts or evidence of symbolic expression does not mean they didn't wield some form of symbolic communication or weren't capable of symbolic expression. Furthermore, evidence of care-giving to the sick and injured goes back even further in the fossil record. Into the deep Middle Palaeolithic, where there's evidence of *Homo erectus* caring for an individual with dental problems who probably incapable of eating properly and would have needed a level of care in advance of what we see in social mammals, even bonobos (Lordkipanidze *et al*, 2005). There is much more evidence of care-giving from pre-sapiens human groups (Spikins *et al*, 2018). So even with no symbolism, these hominins still had compassion. This is another paradox; how did such extensive and consistent levels of care-giving appear so early in the human story when we apparently didn't have language? What was it that drove this care? How could people plan to care for someone without the tool of language?

Well, one explanation is that they did have language, but they just didn't express themselves in any other symbolic way. A good way of testing this would be by looking at the clothing of archaic groups, for example Neanderthals, and looking for artistic intent, or for the presence of symbols. Unfortunately, clothing from ancient groups, especially clothing from archaic, Middle Palaeolithic humans, isn't exactly very common in the archaeological record since it biodegrades. The consensus among researchers is that Neanderthals and other archaic groups may have just worn animal skins with little or no tailoring, if they wore clothes at all (Collard, 2016). Another way we could look for symbolic intent among archaic groups is by looking at the stone tools that these peoples made. Lithic industries (stone tool technology) are one area where the archaeological record is quite rich, even if we go back far into the Middle Palaeolithic. The classic stone hand axe associated with *Homo erectus* is basically a tear-drop shaped large stone usually made from flint. The reason for its shape is that it has been expertly fractured into having an almost symmetrical shape. This could point towards the item being symbolic. Perhaps there was competition between individuals to make the best and most attractive hand-axe (Spikins, 2012). But then, is that true symbolism? It may be artistic in a way, but it's not trying to convey an abstract message to another individual, nor does it represent or symbolize something else.

As we go move further through prehistory from the Middle Palaeolithic lithic technology becomes more complex, and although it is always primarily utilitarian, different styles emerge that differ based on time and locality. This is particularly true of the time around the Middle to Upper Palaeolithic transition, and slightly earlier where, Neanderthals and anatomically modern humans co-existed in Europe and the Middle East for a few thousand years (Shipton *et al*, 2013). At this time, in Europe, we also see an 'explosion' of evidence of symbolic culture (Knight *et al*, 1995), suggesting either cultural traditions rapidly spread through populations in this geographic area, or, a population replaced others through the continent. Africa, holds the origins of symbolism. However, in that continent, evidence of symbolic culture occurs more sporadically, over a longer period of time, from the Middle Stone Age (the African equivalent of the Middle Palaeolithic [these names are based on lithic cultures from the time]) up until the Late Stone Age (Bushozi, 2020). Across the world, by the equivalent of the Upper Palaeolithic, cultures are producing art and decorating utilitarian objects. This is true from Australasia in the South to within the Arctic circle in Siberia, and Patagonia to Alaska.

Mesopotamia is often described as being the 'cradle of civilisation' (Barton, 1929). The period after the Upper Palaeolithic, the Neolithic (or late Stone Age) (a period that started after the Upper Palaeolithic), was allegedly a time when people began to live at higher densities, construct megalithic monuments, and was the period when agriculture began to spread around the world. Mesopotamia, which was prospering during the Neolithic, is a strip of land that is also called the fertile crescent and is located in the Middle East in what is now modern day Iraq, Israel, and Iran. Evidence suggests that the 'fertile crescent' (in Mesopotamia) is where the first villages and cities developed in the Neolithic and Bronze Age (Ur, 2014). By this time, humans in Mesopotamia were developing music, art, and even simple writing. The blossoming cultures of Mesopotamia during the Neolithic and Bronze Age and comparative sophistication compared to virtually all other areas of the world (as far as we know) were probably due to geo-political, and geo-environmental reasons. Luck, in terms of favourable living conditions and rich natural resources also could have helped the early civilisations of Mesopotamia be so successful and well remembered in modern times. It could be argued that the cultures of Neolithic Mesopotamia were the product of Upper Palaeolithic and Late Stone Age societies many social advancements. By 8000 years ago these social changes were starting to be expressed all over the world by different peoples who were becoming sedentary and developing agriculture. The development of early civilisation was bound to happen somewhere in the world around this time. All that was needed was the right geo-political and geo-environmental conditions for a spark to ignite. However, if we follow this argument then we are assuming that there is a defined path of 'simple to complex' in terms of early human culture and social organisation. This could be labelled as ethnocentric and when discussing or comparing any human culture we must maintain cultural relatively as there is no objective truth in terms of culture and cultural practices. Morals and values of different societies are diverse. However, the early civilisations of Mesopotamia certainly are impressive and their achievements awe inspiring.

There are many unanswered questions regarding the origins of symbolic culture but we know that for at least the last 80,000 years (when archaeological evidence of symbolism begins to be found more consistently at sites all over the world) humans have been expressing themselves in metaphorical, abstract ways. This shows that for at least 80,000 years humans have probably been using complex language, since language is, in itself, an abstract, metaphorical symbolic code. However, do the origins of language go back further in time than evidence currently suggests? How closely linked is the emergence of language to symbolic culture? Did complex language emerge rapidly or more slowly over time? The next chapter will discuss this.

CHAPTER 7: THE ORIGINS OF LANGUAGE

The consensus among many groups of archaeologists, evolutionary biologists and anthropologists is that the origins of complex human language is still largely a mystery (Hauser *et al*, 2014). So much so that debating the origins of language was, for a long time around the 'enlightenment' era, banned at intellectual meetings (Szamado, 2004). We use a variety of vocally produced sounds to create words and sentences that have a near infinite potential meaning. Through this we can convey complex messages, about the present, about the past, and about the future to one another which allows modern humans to communicate about imminent dangers, to share nostalgia about what has been, and to plan complex strategies to protect ourselves and others from future events. However, our lineage hasn't always had the benefits of complex language. Some argue that language only emerged with the birth of symbolism, and we only see consistent evidence of symbolic behaviour in the last 80,000 years or so. This would imply that language emerged relatively abruptly, although it's emergence may alternatively have been in incremental steps.

As language is built on abstract code (some words refer to concepts which only exist in our minds, are arbitrary and representational) it is easy how one could argue that it only appeared with, or at the same time as symbolic culture. However, it is also possible that some forms of symbolism go back far further. This alternative is that proto-languages were used by our early Middle Palaeolithic ancestors. There is a problem with this second explanation though. Chomsky (1996) argues "true language, via the emergence of syntax, was a catastrophic event, occurring within the first few generations of *Homo sapiens sapiens* (Chomsky, 1996)". So according to some linguists, syntax, the ability to create well formed words and phrases to create meaningful sentences, is unlikely to have evolved slowly, but must have appeared rapidly, in a revolution perhaps. It may have been that those first humans with the tool of language were the first humans to have, what we would define it as now, a psychotic episode. In other words, a few individuals may have begun to think metaphorically in symbols, and this then was selected for because of some social advantage, and spread throughout the population. This way of thinking, using symbols to rationalise the past, present, and future, must have been useful since it survived and spread to different populations around the world.

Is symbolic expression analogous to language? It is probably true that evidence of language is harder to detect than evidence of symbolism (becuase vocal or gestural language leaves no archaeological trace) but if ancient *Hominins* from the Middle Palaeolithic did create imaginary worlds and tell stories (a universal among the *Homo sapiens* groups that have been studied via ethnographic studies) then is there a way of collecting evidence to support this theory? We have sporadic evidence from the very distant past. Such as a sea shell engraved with zig-zag lines from over 500,000 years ago (Joordens *et al*, 2015) worked on by a *Homo erectus* individual from Java. Another case is the Makapansgat pebble which was carried for potentially symbolic reasons (Davidson, 2020). Speaking of *Homo erectus*, the hobbit men (*Homo floresiensis*) from Flores, who may have evolved from Asian *Homo erectus*, could allegedly use language to some capacity (Brumm *et al*, 2006). Most of the evidence we have of this are the stories that have been passed down through the generations by the people of Flores (Brumm *et al*, 2006), and some tools that may have belonged to them. The origins of syntax is a major mystery for understanding the origins of language currently (Deacon, 1997). Perhaps, the answer will be such a shock to Western Civilisation that it'll cause a paradigm shift, and a new beginning will start. Well developed theory of mind and orders of intentionality, which are thought to be key to how modern humans experience the world and interact socially (de Waal, 2006), may be important because in order to avoid social deception we must have a good idea of the intentions of others, while also knowing how our own actions have consequences (a result of theory of mind).

There are various cognitive models which try to help us to work out how the material evidence left behind by archaic hunter-gatherers can be used to infer how the people that produced that evidence thought (Malafouris, 2019). Whether they rationalised the past, present, and future through using symbols like we do, or as the lack of symbolic evidence implies, the argument that some archaic hunter-gatherers did not think representationally and metaphorically, and instead understood the world as they saw it may have validity. As in these hominins didn't have the inclination to imagine various possible scenarios, the chain of thoughts of which may have been the first step in creating complex plans which would have massively aided chances of survival. If this wasn't the case and they saw reality 'as it is' (similarly to animals that lack symbolism and theory of mind) then their forms of communication must have been very interesting to observe, because it would be a analogy of an anatomically modern human behaving and communicating similarly to a different mammal in a social, and natural context. With theory of mind we can understand events through time, and it is because of theory of mind that language has utility. Our well developed theory of mind may have been key to the evolution of language because of this.

The utility, efficiency, and adaptability of complex human language is exemplified through writing. A relatively new function of our symbolic communication system that has only emerged in the last 5000 years (Schmandt-Besserat, 1978). Writing probably first emerged simply as a way to keep records of traded items, tax collection, or for some other reason related to the economy of early large settlements (Schmandt-Besserat, 1978). However, it has become a way for *Homo sapiens* to communicate across time because unlike the spoken word, writing has the potential to be preserved for thousands of years or potentially longer (now with the internet and digital technology another leap forward has been made). The proliferation of writing has allowed us to 'hear' peoples from thousands of years ago. Whether through stories, scientific achievements, or philosophical perspectives writing has allowed modern people to connect with our ancestors from hundreds and thousands of years ago. As writing is simply language written down it uses symbols and representational signs to communicate with others the same as the spoken word. Cumulative culture (Caldwell, 2009) then gains speed and impact. Since writing has emerged it has allowed peoples to advance intellectually and technologically based on the errors and successes of others in the past. However, writing also brings with it a feeling of privilege and superiority in terms of history being more important than prehistory because writing gives us an understanding of what an era was like, whereas prehistory wasn't recorded. In prehistory great technological advancements were made that are still mysterious to us today such as the creation of megalithic structures (Stonehenge, for example) and perhaps even greater and more marvellous events occurred such as the birth of religion, or music, that we don't know of through recorded history. Writing is an interesting and wondrous creation but it was not the origin of language or symbolic culture. It is a product of language and symbolic culture being expressed and blossoming into new fields at a later stage than the actual origin of symbolism.

Those who follow Chomsky's view are known as saltationists, and argue that language emerged 'catastrophically' (Hauser *et al*, 2002). In other words a change occurred in one or more individuals that allowed lexical items to be recursively combined, thus creating proper syntax (Currie, 2019). Saltationists argue that prior to this, symbols were used in thought, but never expressed in 'conspecific language' (Hauser *et al*, 2002). So saltationists think that hominins were thinking to themselves in abstract, representational (and perhaps psychotic) ways, before hominins started actually using this symbolic ability to communicate with one another. Gradualists on the other hand, as the name suggests (Hauser *et al*, 2002) believe that there were 'protolanguages' which preceded complex human language that created syntax. There is evidence for both points of view. For example, for followers of Chomsky, consistent evidence of symbolism doesn't appear until relatively recently in the archaeological record. Symbolism, as has been argued, is analogous to language, to a degree, so this implies that there was some major hurdle that had been crossed relatively recently in regards to the expression of symbols and the use of symbols as a communication system. On the other hand, for gradualists, hand axes of the Archeulean lithic industry associated with homo erectus are highly symmetrical and obviously great care, effort, and risk were involved in producing them so this implies cultural richness and learning through generations which would definitely have been aided with some form of proto-language. Also, 500,000 years ago *Homo heidelbergensis* were planning and executing risky, dangerous hunts of large wild animals (Voormolen, 2008), this certainly would have been more likely to succeed with a proto-language to help planning and communication. What we can know about the origins of language is that at some point in hominin evolution, full symbolic communication between conspecifics was achieved and eventually spread to all hominin populations.

Fundamentally, symbols are used by modern humans to convey representational meaning and display allegiance to common beliefs or values shared by individuals and groups. For example, in much of the world, football (soccer) crests are emblems which represent a certain team from a certain place, and those that recognise and follow that team will easily recognise the crest. This is one way people have used symbols. *Homo sapiens* also use and have used symbols to more generally communicate. A lavatory sign is easily recognisable to most people as a place where one can use the toilet and freshen up. It contains no other explicit or implicit meaning, which is more arguable when we are talking about a football crest. The design of a football crest may refer to the clubs history, a rivalry, or a myth surrounding the team's past and legacy. Writing, in terms of the words that you are reading now, contain no special abstract or implicit meaning. The script used is just there so that letters can be put together to create words and sentences (with an almost infinite amount of combinations, and hence meaning being possible) which convey a message. Therefore, symbols don't have to be abstract. They can just convey a utilitarian meaning, with no symbolic, mystical, or abstract undertones.

However, were the origins of symbolic communication for a utilitarian purpose or a mystical, ritual, or abstract one? Chimpanzees can be taught to use symbols (Meddin, 1979), but once in an environment only with their conspecifics they very rarely choose this option (Boesch, 1991). What does this imply? That it is easier to not use language or another form of symbolic communication when interacting but to use physical coercion or physical threats in order to dominate your peers? Human social complexity means that if that occurred in any human society, the perpetrator would be immediately ostracised or punished severely. Ethnographic studies of egalitarian hunter-gatherers is testament to this (Cashdan, 1980). Another testament to this is the theory that we are 'self-domesticated' (Theofanopoulou, 2017). To recap, those individuals who expressed tolerance rather than reactive aggression, and perhaps were more caring or artistic, were consistently selected for over time (through natural selection) and our species gradually became less physically 'robust'. This process may have resulted in large, overly aggressive, dominant males being less successful in our immediate ancestors than in those of our close cousin, the Chimpanzee.

Perhaps, the start of symbolism was neither for utilitarian nor ritual use, but as a way to rationalise or understand events, environments, processes in our heads. The block of red ochre, with patterns engraved in it, from basically the start of the emergence of anatomically modern humans over 100,000 years ago (Watts, 2002) may indicate something interesting. Because this piece of art isn't embedded in a richer cultural context, such as being associated with an elaborate burial, or a parietal art (cave art) diorama, it may mean that the individual who created it was simply bored, and felt the impulse to engrave this piece of ochre. This indicates that the thought processes of this individual at the time, were abstract and symbolic. Similarly to all modern humans today. The individual may have subconsciously created the pattern to help rationalise the thoughts that were going through their own mind.

This interpretation is based on cognitive archaeology models and theories. Cognitive archaeology's goal is to understand the thought processes of ancient peoples, based on the material evidence that is left behind. Malafouris (2019) is a key proponent of material engagement theory, which postulates that the human mind is made up of not just what goes on inside the skull, but also our hands and extending from this, the tools we use to create material culture. In other words the material products that we produce are an extension of our cognition. This theory can be applied to all member of our genus. Since we started creating simple tools. However, it is since symbolism emerged and then proliferated that the theory becomes more useful, since the more artistic, rich material culture that is produced, the more material cognitive archaeologists have to work with to infer cognition.

Linguistics and the Origin of the Symbol

Another perspective we could take in investigating the origins of symbolism is to look at work done by Linguists. While we've covered some of the linguist, Noam Chomsky's ideas, this section will go back to basics in terms of linguistic ideas. This primarily covers the types of signs that exist, including symbols, in human language (Short, 2007). There are icons, which resemble what they represent (a painting of an apple is iconic of an actual apple). There are also indices, which indicate something (a weather vane indicates the direction of the wind). Finally, there are symbols which are arbitrary (have no relation to what they represent) and are only given meaning by a group of people agreeing on what they mean. For example the word 'chair' isn't iconic and doesn't indicate what it represents. It is arbitrary (Short, 2007).

The thinking goes, that symbols appeared last in human evolutionary history. The icon and indices appeared first and gradually composite signs emerged combining these elements, until as theory of mind and orders of intentionality grew in human groups, we were able to develop complex symbols.

It must have been a huge jump for our species and genus to develop composite signs, whether in gestural form or spoken. I say this because it had never happened before as humans are unique in the animal kingdom in using complex symbolic language (as far as we know). Linguistics is amazing as it can reconstruct the origins of fairy tales (some can be traced back over 5000 years) (Zipes, 2011), so, of course it is a good resource to use to understand human symbolism. Again, it seems that because symbols are arbitrary and only given meaning through human consensus, we are all creating and involved with each others, symbolic worlds. To add to this, I wonder if the moral line between right and wrong is easily crossed in our symbolic worlds, or if this line is subjective and there is no objective 'good'?

Understanding the origins of the symbol and symbolic culture more generally is a cross-disciplinary effort, and interaction between different fields would help find some answers. However, it is very difficult to reconstruct the past, and while theories are interesting and make you think, we might never be able to go back and actually see for ourselves (although computer simulations could give some answers). We as humans like to categorize and label things as something, when in reality, is that a definition of something 'concrete'? If we look at the labelling of people, this could be a weight around someone's shoulders that limits their potential.

The symbolic world that humans have broken into might give endless possibilities, as with psychosis/schizophrenia anything could be possible, but does this mean that we also put labels on people that are detrimental? Is there a flow of energy where we have to do wrong in order to gain power? Hopefully this isn't the case, and as with karmatic flow good is rewarded. Is there freedom from this cycle though (or is the whole concept not real)?

The west and its traps have for hundreds of years tried to control it's populace with covert control. The groups that are oppressed may change over time, but it seems to be a constant that different groups will become oppressed. It is part of human psychology to need an enemy and an ally. However, we can all be rational and altruistic. We have evolved altruism because of inclusive fitness. Those that we help and are kind to may also reciprocate that kindness one day. We shouldn't be kind for this reason though, and more importantly shouldn't take evolutionary theory as a model for how we act to one another. Science can be harsh, cold, and isn't a great template to use to model society. There are some things we can't explain. We all have the power as individuals to make the world a better place. Let's use the positivity of our good experiences to spread compassion, and make the world a better place.

This chapter has looked at the origins of language. Icons, indices, and symbols are all signs, and symbols are uniquely used regularly by humans. The symbol can be defined as an arbitrary sign, that is given meaning through consensus. Language probably emerged either vocally or gesturally, and slowly developed over time. It is closely linked to other symbolic expression such as art and music. Writing is a very recent addition to our repertoire but it allows us to mentally time travel to thousands of years in the past. Symbolic culture first started in Africa sporadically and perhaps based on population density, geopolitical circumstances, and maybe selection pressures, had bursts of production in different parts of the world. Africa was the birthplace of anatomically modern humans and also human culture. Talking about race within the field of Anthropology is often contentious, but the great genetic diversity of humanity (particularly within Africa) increases the chances of diversity in thought and ways of social organisation. This helped our species survive and achieve success in the past, because greater diversity means a greater chance to be able to fill different ecological (and social) niches, and may also help our species in the future. The next chapter discusses race.

CHAPTER 8: RACE, LABELLING, AND DETERMINISM

All humans alive today are *Homo sapiens sapiens*. We have a very recent common ancestor of about 100,000 years (Chang, 1999). Most of the great leaps forward in our evolution happened in Africa, including the development of symbolism. By this I mean that art, music, and religion first emerged in Africa with archaic *Homo sapiens*. These people were more robust than us today, and with them symbolism spread to Europe and Asia. It is hypothesized that Eurasian Neanderthals became more symbolic due to contact with early *Homo sapiens*. The freedom of thought and access to the symbolic realm brought with the ability to ponder existence, express ourselves in thousands of ways, and communicate with each other on deeper levels than ever before. This is how we all went crazy (from a certain perspective).

As humans, we like to categorize objects and groups into definable units. This obsession with discontinuous categories has led us to define humanity based on skin colour. However, is this actually taxonomically accurate? The inequality based on perceived differences is not only un-scientific, but offensive and built on the oppression that past generations have had to endure.

Humanity's love for discontinuous categories can be seen clearly with how we define extinct hominin 'species'. *Homo rudolfensis*, *Homo Heidelbergensis*, and *Homo Neanderthalensis* could all breed and produce fertile offspring with each other (Dalton, 2010) so they're not technically different species (based on most technical definitions of species). It is just part of the human condition that we like to argue for differences because these groups didn't have civilisation despite their complex tool-making, subsistence strategies and culture.

Across species, genetic diversity can be used as an indicator to how likely a species is to adapt to environmental change. The greater the genetic diversity, the more likely it is to adapt to change. In modern humans, we find more genetic diversity within populations than between them (Lewontin, 1972). This indicates that there are more phenotypic differences within a race than between races. What this implies is that the great atrocities of eugenics were based on falsehood and are pseudo-scientific. We are more closely related to any other human, than a Chimpanzee is from a conspecific that lives on the opposite side of the mountain that splits its population.

Furthermore, we all have a very recent common ancestor. Every modern human alive today can trace back our last common ancestor to around 100,000-200,000 years ago (Chang, 1999). This is the blink of an eye evolutionarily speaking. As a comparison tigers (Panthera tigris) branched off from other big cats 3.2 million years ago. This is one line of evidence to suggest racism is based on falsehood, and power that is gained from geo-political and geo-economical advantages (in terms of Europeans thinking they are superior). In other words, race is primarily a social construct and it is race, biology, and culture which all combine to create the racial differences that exist (Gravlee, 2009).

From a more sociological approach, racial inequality is a result of institutional racism and the systems in place, which disadvantage certain ethnic minorities (Gimenez, 2018). In other words, inequality is a class issue, not a race issue. For example, if two young people get into trouble with the police, but one has parents who are in well paid, prestigious jobs (such as teaching), compared to the other, who's parents are cleaners, they may be more likely to talk their way out of the trouble with the police. This is known as 'elaborate code' vs 'restricted code' (Brannigan et al, 1976). This implies class and money are the main factors which influence incarceration and achievement. IQ tests may also be culturally biased, and the compartmentalism of the brain may show that we are all products of our environment, acquiring different strengths based on our experiences. Nature vs nurture is a theme that I keep going back to and it's an important argument in regards to whether someone will reach their 'full potential'. If a child is raised with no human contact the resulting person is highly damaged and unable to interact in the same way as a person who has a caring primary and secondary socialisation.

People who are to the right of the political spectrum often emphasise the power of nature, while those on the left often emphasise the power of nurture. A good example of this is how fascists of 20th century Europe believed in the determinism of our racial origins (genetic origins) as a predictor of how that individual would operate in society. Human experience and life journeys may transcend genetic predispositions, as the impact of nurture overrides and is more powerful than the limited genetic differences between us. There is greater diversity within 'races' than between them. Hence 'race' is a social construct.

Power can be maintained by states or tyrants through a divide and conquer approach. The western media and other state apparatus are at the moment projecting the ideas of some young people that there are multiple ways to define ones gender identity and sexual orientation. While this may have its benefits for some, in terms of personal growth and acceptance, the elevation of these ideas is also an underhanded way of dividing us so we are easier to control. Coming back to racial labels and stereotypes, these multiple categories have a similar function. When people are divided and fighting amongst themselves they are easier to control.

It makes one wonder if there are sociologists employed by governments to assess what stance the institutional state apparatus and repressive state apparatus such as the media and the police should take on certain issues. For example, with movements based on race, class, or sexuality (while they may also be good in achieving positive goals) state sociologists could argue for the movement to be supported but put emphasis on the difference between people in order to divide the population because that makes them easier to control. This makes one think if there is some truth to the myth of secret societies, like the illuminati. Media, but especially politicians lie constantly about many different issues. It is hard to get someone's honest opinion about important issues in this world. Once labels become self-fulfilling prophecies and lead to someone's corruption (in some form or other) other people tend not to speak to them in an honest, genuine way.

Even labels that we give ourselves, positively or negatively can turn into self fulfilling prophecies, and with mental illness it can be difficult to control what goes through the mind. Nasty individuals abusing these vulnerable people doesn't help the thought patterns of these 'mentally ill' individuals. Meditation can help to ease this and control thoughts/ Having a hyperactive mind is not healthy. Whether drugs, or environment can help us to attain peace that is a path that potentially could work. Labelling someone as different divides people and so maybe basing identity on race or sexuality isn't the best tactic, we should all take.

Linked to labelling, genetic determinism is the idea that because your genes are of a certain type you are due a certain fate. This is a right-wing concept. We are all free to explore what life can offer and fight for what we can be. To go even further, social and identity determinism is an insult to our human rights, and is basically fascism. To curtail someone's human rights, and life chances, because of something that they have no power over, is, by definition, fascism. Furthermore, due to labelling in western society, people are losing their human rights, and their prospects and aspirations are being destroyed. It's a myth that there is equality for people in western society, when you grow to be an adult, if you have the right viewpoint, you'll see the dark underworld of the 'equal' society that we live in. This can be exemplified by looking at individual people's life stories that you will come across in said underworld.

What about determinism based on perceived sexuality? Like racial, or class determinism this can limit someone's potential. For example, a labourer's son may be more likely to be given a label as someone who will definitely be best at manual jobs and definitely shouldn't be encouraged to study maths, or science. Or, someone who isn't white being thought of as more likely to get into trouble so special measures are put on them in school which could be detrimental to their potential academic achievement. However, with sexuality, I think that it is even more disgustingly fascist than with the previous two examples. Firstly, the whole system of labelling someone as 'gay' is inherently homophobic in itself. The label has negative connotations associated with it. We're all people, we shouldn't be forced to cut 50% of the people that we could potentially be with. Imagine, that child becomes introverted because of this label, and can't express themselves. Maybe without this label they'd feel like they could. Also, what if the definition, and the label that the child (or adult) is given is wrong. Do you really know what gender someone is attracted to depending on their mannerisms or if they are masculine or effeminate (or dominant or submissive)? If western society is seriously trying to say that 'gay' children can be abused by the sex that they are attracted to then it is disgustingly abhorrent on multiple levels.

Humanity is obsessed with categorizing groups into definable units as can be seen with our classifications of archaic groups. Racism is also a falsehood based on a pseudo-science (eugenics) that is born out of geo-economic differences. Biologically, there is more difference within races than between them which points towards races being a social construct. Finally, sociology can shed light on racism and show that identity is based on many features and while skin colour may be one, it is certainly not the only one.

The great diversity of humans across the planet allows for many natural socio-political experiments in terms of social organisation because of great genetic diversity. The robustness of this diversity and the individuals and groups that it could produce depends on who you ask. This is based on where people land on the nature/nurture spectrum in terms of the adult human product. In the modern world there is some conscious choice in form of social organisation one finds themselves in. However, we don't all have the choice to live in a Kibbutz, or join a self-sufficient vegan community in the countryside somewhere. The diversity of Africa was key in the natural social experiments of the Middle/ Late Stone Age and the Middle/Upper Palaeolithic. What I mean is the social experiments in terms of evolution, not conscious experiments. These different modes of social organisation occurred in phases across thousands of years and eventually (possibly due to the emergence of hierarchy and then inequality) led to blossoming of symbolic culture.

CHAPTER 9: IDEAS CHANGE: ARE WE CLOSER TO THE TRUTH?

In this book I have tried to lay out some theories which explain the evolution of symbolism, including the emergence of language. I've also discussed some of the similarities that we share with other intelligent social creatures on this earth in relation to communication between conspecifics. Finally, I've tried to persuade the reader of how human diversity is so valuable, yet differences are often targeted by those wishing to control in order to divide us. As we look back through history, we see many atrocities that have been committed, and often we like to think that now things are safer, less violent, and in general it is a nicer world to live in. However, is this really true? Is it right to be optimistic or pessimistic about humanity in the 21st century? Are we blinded to blatant abuse and injustices going on right in front of our eyes? Are some of the attitudes and beliefs which lead to unkindness and abuse something society has always had since the dawn of humanity or since the dawn of big cities? Does it go further back to the dawn of the *Homo* genus itself? Or is this an illusion created by our tiny, individualistic experiences and, on a grand scale, things have actually got better in terms of suffering as the media like to imply?

The future of humanity is uncertain. Some of us want to venture to the stars and expand human civilisation to worlds beyond our own. This is an ambitious, beautiful target but will we destroy ourselves in this rush? Would it not be better to make the world we are living in fairer, and a more just place for all of us? We could build an eco-friendly civilisation where we grow crops but do not farm animals in order to end the suffering of livestock. We could re-wild and breed animals similar to aurochs and mammoth and live in peace with nature. Whole ecosystems could be re-created that blend seamlessly in with human settlments. We could build grand eco-friendly communities with grass, trees, and bushes all around, and species that are good for the ecosystem and environment living around us. We are clever enough for this type of society to take shape. There are great minds living on this earth and through our great ability to plan an innovate we could make this work.

However, the greed of capitalism is a big obstacle to any potential socialist or anarchist utopia. The corporations that currently control the world are primarily interested in one thing, profit. They will cut corners to gain small advantages. This system is ironically similar to natural selection in that the 'weak' lose and are killed off. However, despite the perils of organised religion such as totalitarianism, witch-hunts, and abuse, the optimistic idea that humans can change the ruthlessness of nature and in turn the capitalist system could have some truth and this idea is borrowed from religious values and principles.

.

In the mid 20th Century, communism was putting up a fight against capitalism and imperialism but it now seems that capitalism has won. The world is so globalised, if a revolution is spawned, maybe it'll spread like wildfire. Or, maybe it won't because the fascist systems of society and governments have won. Maybe we're so interconnected that we are almost one unit, and factions appearing aren't possible anymore. We have never been as interconnected to so many people ever before. However, in the past we were closer, more equal, and more intimate with fellow humans than we are now, can we ever return to that kind of intimate seclusion?

In the deep past, we had a foundation of cooperation, egalitarianism, intelligence and tool making. At some point, probably before some of the features that I've just mentioned emerged, a mental spark was lit in a few hominin individuals, and that allowed access to imagination, and the symbolic realm. From this development we soon developed mythical story telling and could communicate and express ourselves in an almost infinite range of ways. This 'spark' was the most miraculous event in the history of the hominin line and could have been aided with the help of psychedelic drugs or strong social bonds towards one another. This was humanity's most crazy moment. As the contrast of opinion and world views show, we all have a different way of seeing reality, and it is probably easy to wander off the mental or cognitive path of normality. However, being open minded about the views of others, and helping rather than condemning, could lead all of us to lead lives with less suffering. Language and symbolism have opened so many imaginative worlds to humans, including the ability to more easily define right or wrong (maybe), so if we take the time to actually think, we can help to struggle for a world that is more based on justice.

Humanity could enjoy it's position in this life, appreciate the great understanding we have of the natural world and our progress in mastering evolutionary mechanisms, engineering principles, and religious teachings, and not think of life as a competition. Currently, our capitalist economy encourages working against each other, and rewards taking advantage of other's weaknesses, in a business sense. While environmental, and socialist initiatives are being implemented, will this be enough to stop the environmental destruction that is underway? We are approaching 10 billion people living on earth and we do not have the resources to maintain this population explosion. Our economic systems are not built to resolve the impending environmental crisis. We could go back to basics, have smaller classes in education systems, learn bush craft, plant forests, and re-wild by reintroducing extinct species. We should strive to make our environments as green as possible.

It's a great feeling to breath in the high oxygen content of the air in forests and wild places. If we could have plants growing n all our outdoor spaces, and fill the air with oxygen, it would be a positive thing for the atmosphere and environment not only in our own countries but around the world. We have the capability to make these changes but our economic systems are pushing humanity in the other direction, to competition and destruction.

In terms of reaching the stars, the 'great filter' concept (Jiang, 2021) is the idea that, because we haven't found alien life, there may be an evolutionary step which is very difficult to achieve for life. It could be the birth of single celled organisms, the birth of multicellular organisms, or the evolution of intelligence. The other option is that the 'great filter' is ahead of us, and humanity will most probably be destroyed at that step in evolution of civilisation. However, because we have found no alien life it is possible the great filter is behind us as we have no evidence of life evolving independently into single celled, multi-celled, or intelligent life. It may be that once life evolves into advanced civilisations it destroys itself with nuclear bombs. However, nothing is certain and we must always have hope for the future. Part of human access to the symbolic realm is that we can have different kinds of hope for the future.

The fact that there are multiple perspectives that we can take on the current moral and social state of humans across the world in itself shows that there is no one truth. The fact that we all think slightly differently and drugs impact our perception of reality may show that there is no concrete truth? As there may be no one truth why is one viewpoint more accurate than another? Is there in fact an objective reality? This further backs up my argument that we are all psychotic apes because if there is no reality, what control do we have over events, and how is one person's world view more valid than another's if there is no objective truth?

With over 9 billion people on earth, if one person does see a deep injustice that they feel needs to be changed is it possible for them to make a difference? I'd argue that, yes, it is possible. Whether that's through obtaining upwards social mobility and becoming influential in important circles, making a social statement through your behaviour, or sitting down and writing a book. If the animal kingdom can display such a diversity of forms of social organisation why do humans have to be any different. Be the change that you want in this world.

References:

Aiello, Leslie C., and Peter Wheeler. "The Expensive-Tissue Hypothesis: The Brain and the Digestive System in Human and Primate Evolution." *Current Anthropology*, vol. 36, no. 2, 1995, pp. 199–221. *JSTOR*, http://www.jstor.org/stable/2744104. Accessed 13 Mar. 2023.

Aldhebiani AY. Species concept and speciation. Saudi J Biol Sci. 2018 Mar;25(3):437-440. doi: 10.1016/j.sjbs.2017.04.013. Epub 2017 May 3. PMID: 29686507; PMCID: PMC5910646.

Amodio, Piero & Boeckle, Markus & Schnell, Alexandra & Ostojic, Ljerka & Fiorito, Graziano & Clayton, Nicola. (2018). Grow Smart and Die Young: Why Did Cephalopods Evolve Intelligence?. Trends in Ecology & Evolution. 34. 10.1016/j.tree.2018.10.010.

Angelbeck, Bill. *Canadian Journal of Archaeology / Journal Canadien d'Archéologie*, vol. 30, no. 1, 2006, pp. 101–04. *JSTOR*, http://www.jstor.org/stable/41103554. Accessed 17 Mar. 2023.

Anton S.C, Snodgrass, J.J. 2012. *Origins and Evolution of Genus Homo*: *New Perspectives*.

Current Anthropology 2012 53:S6, S479-S496.

Antón, Susan. (2003). Natural History of Homo erectus. American journal of physical anthropology. Suppl 37. 126-70. 10.1002/ajpa.10399.

Argue, Debbie & Donlon, Denise & Groves, Colin & Wright, Richard. (2006). Homo floresiensis: Microcephalic, Pygmoid, Australopithecus, or Homo?. Journal of human evolution. 51. 360-74. 10.1016/j.jhevol.2006.04.013.

Armit, Ian & Knsel, Chris & Robb, John & Schulting, Rick. (2006). Warfare and Violence in Prehistoric Europe: an Introduction. Journal of Conflict Archaeology. 2. 1-11. 10.1163/157407706778942349.

Baglione, Vittorio & Canestrari, Daniela. (2016). Carrion crows: Family living and helping in a flexible social system. 10.1017/

CBO9781107338357.007.

Barton, George A. "The Origins of Civilization in Africa and Mesopotamia, Their Relative Antiquity and Interplay." *Proceedings of the American Philosophical Society*, vol. 68, no. 4, 1929, pp. 303–12. *JSTOR*, http://www.jstor.org/stable/984342. Accessed 14 Mar. 2023.

Bates, Lucy & Lee, Phyllis & Njiraini, Norah & Poole, Joyce & Sayialel, Katito & Sayialel, Soila & Moss, Cynthia & Byrne, Richard. (2008). Do Elephants Show Empathy?. Journal of Consciousness Studies. 15.

Bednarik, Robert. (2008). The domestication of humans. Anthropologie. 46. 1-17.

Ben-Dor M, Gopher A, Hershkovitz I, Barkai R. Man the fat hunter: the demise of Homo erectus and the emergence of a new hominin lineage in the Middle Pleistocene (ca. 400 kyr) Levant. PLoS One. 2011;6(12):e28689. doi: 10.1371/journal.pone.0028689. Epub 2011 Dec 9. PMID: 22174868; PMCID: PMC3235142.

Bentley-Condit, Vicki K., and E. O. Smith. "Animal Tool Use: Current Definitions and an Updated Comprehensive Catalog." *Behaviour*, vol. 147, no. 2, 2010, pp. 185–221. *JSTOR*, http://www.jstor.org/stable/40599646. Accessed 14 Mar. 2023.

Bocquet-Appel, Jean-Pierre & Degioanni, Anna. (2013). Neanderthal Demographic Estimates. Current Anthropology. NS 8. S202-S213. 10.1086/673725.

Boesch, C. Symbolic communication in wild chimpanzees?. *Hum. Evol.* **6**, 81–89 (1991). https://doi.org/10.1007/BF02435610

Borrego, Natalia. (2016). Social carnivores outperform asocial carnivores on an innovative problem. Animal Behaviour. in press. 10.1016/j.anbehav.2016.01.013.

Bickerton D, Szathmáry E. Confrontational scavenging as a possible source for language and cooperation. BMC Evol Biol. 2011 Sep 20;11:261. doi: 10.1186/1471-2148-11-261. PMID: 21933413; PMCID: PMC3188516.

Brandl, J.L. The puzzle of mirror self-recognition. *Phenom Cogn Sci* **17**, 279–304 (2018). https://doi.org/10.1007/s11097-016-9486-7

Brannigan, Augustine, and D. Lawrence Wieder. "Language and Social Reality: The Case of Telling the Convict Code." Contemporary Sociology, vol. 5, no. 3, 1976, p. 349., doi:10.2307/2064132.

Brecht, Katharina & Wagener, Lysann & Ostojic, Ljerka & Clayton, Nicola & Nieder, Andreas. (2017). Comparing the face inversion effect in crows and humans. Journal of Comparative Physiology A. 203. 10.1007/s00359-017-1211-7.

Breuer, Thomas. (2002). Distribution, Feeding Ecology and Conservation of the African Wild Dog (Lycaon pictus) in Northern Cameroon.

Brown, A. (2011). *The rise and fall of communism*. ECCO.

Brown, Heather. (2013). Marx on Gender and the Family: A Critical Study. 10.1163/9789004230484.

Brumm, Adam & Aziz, Fachroel & Van den Bergh, Gert & Morwood, Michael & Moore, Mark & Kurniawan, Iwan & Hobbs, Douglas & Fullagar, Richard. (2006). Early stone technology on Flores and its implications for Homo floresiensis. Nature. 441. 624-8. 10.1038/nature04618.

Burdukiewicz, Jan. (2014). The origin of symbolic behavior of Middle Palaeolithic humans: Recent controversies. Quaternary International. 326-327. 398–405. 10.1016/j.quaint.2013.08.042.

Bushozi, P. M. (2020). Middle and Later Stone Age Symbolism, *Utafiti*, *15*(1), 1-27. doi: https://doi.org/10.1163/26836408-15010020

Caldwell, Christine A., and Ailsa E. Millen. "Social Learning Mechanisms and Cumulative Cultural Evolution: Is Imitation Necessary?" *Psychological Science*, vol. 20, no. 12, 2009, pp. 1478–83. *JSTOR*, http://www.jstor.org/stable/40575214. Accessed 16 Mar. 2023.

Capasso, Luigi & Michetti, E. & D'Anastasio, Ruggero. (2008). Homo erectus hyoid bone and the origin of speech. Coll. Antropol.. 32. 315-319.

Cashdan, Elizabeth A. "Egalitarianism among Hunters and Gatherers." *American Anthropologist*, vol. 82, no. 1, 1980, pp. 116–20.*JSTOR*, http://www.jstor.org/stable/676134. Accessed 16 Mar. 2023.

Chomsky, N, 1996. *Powers and Prospects. Reflections on human nature and the social order*. London: Pluto Press, p 30.

Cissewski, Julia & Luncz, Lydia. (2021). Symbolic Signal Use in Wild Chimpanzee Gestural Communication?: A Theoretical Framework. Frontiers in Psychology. 12. 10.3389/fpsyg.2021.718414.

Chang, Joseph T. "Recent Common Ancestors of All Present-Day Individuals." Advances in Applied Probability, vol. 31, no. 4, 1999, pp. 1002–1026., doi:10.1239/aap/1029955256.

Chappell, Jackie. (2006). Living with the Trickster: Crows, Ravens, and Human Culture. PLoS Biology. 4. 10.1371/journal.pbio.0040014.

Clement AF, Hillson SW, Aiello LC. Tooth wear, Neanderthal facial morphology and the anterior dental loading hypothesis. J Hum Evol. 2012 Mar;62(3):367-76. doi: 10.1016/j.jhevol.2011.11.014. Epub 2012 Feb 17. PMID: 22341317.

Collard, M., Tarle, L., Sandgathe, D., & Allan, A. (2016). Faunal evidence for a difference in clothing use between Neanderthals and early modern humans in Europe. *Journal of Anthropological Archaeology*, *44*(Part B), 235-246. https://doi.org/10.1016/j.jaa.2016.07.010

Conard, Nicholas & Serangeli, Jordi & Bigga, Gerlinde & Rots, Veerle. (2020). A 300,000-year-old throwing stick from Schöningen, northern

Germany, documents the evolution of human hunting. Nature Ecology & Evolution. 4. 1-4. 10.1038/s41559-020-1139-0.

Conard, Nicholas & Serangeli, Jordi & Böhner, Utz & Starkovich, Britt & Miller, Christopher & Urban, Brigitte & Kolfschoten, Thijs. (2015). Excavations at Schöningen and paradigm shifts in human evolution. Journal of Human Evolution. 89. 10.1016/j.jhevol.2015.10.003.

Conard, Nicholas. (2006). Changing Views of the Relationship between Neanderthals and Modern Humans..

Coolidge, Frederick & Wynn, Thomas. (2009). Homo heidelbergensis and the Beginnings of Modern Thinking. 10.1002/9781444308297.ch9.

Corvinus, Gudrun. (2004). Homo erectus in East and Southeast Asia, and the questions of the age of the species and its association with stone artifacts, with special attention to handaxe-like tools. Quaternary International. 117. 141-151. 10.1016/S1040-6182(03)00124-1.

Creel, Scott & Creel, Nancy & Mills, Michael & Monfort, Steven. (1997). Rank and reproduction in cooperatively breeding African wild dogs: Behavioral and endorcrine correlates.. Behavioral Ecology. 8. 298-306. 10.1093/beheco/8.3.298.

Cunha, Gerald & Wang, Yuzhuo & Place, Ned & Liu, Wenhui & Baskin, Larry & Glickman, Stephen. (2003). Urogenital system of the spotted hyena (Crocuta crocuta Erxleben): A functional histological study. Journal of morphology. 256. 205-18. 10.1002/jmor.10085.

Currie, A, Killin, A. From things to thinking: Cognitive archaeology. *Mind Lang*. 2019; 34: 263– 279. https://doi.org/10.1111/mila.12230

Dalton, Rex. "Neanderthals May Have Interbred with Humans." Nature, 2010, doi:10.1038/news.2010.194.

Dannemann M, Kelso J. The Contribution of Neanderthals to Phenotypic Variation in Modern Humans. Am J Hum Genet. 2017 Oct 5;101(4):578-589. doi: 10.1016/j.ajhg.2017.09.010. PMID: 28985494; PMCID: PMC5630192

Davidson, Iain. (2020). Marks, Pictures and Art: Their Contribution to Revolutions in Communication. Journal of Archaeological Method and Theory. 27. 10.1007/s10816-020-09472-9.

Deacon, Terrence William. The Symbolic Species : the Co-Evolution of Language and the Brain. New York :W.W. Norton, 1997.

Dean, Lewis & Vale, Gill & Laland, Kevin & Flynn, Emma & Kendal, Rachel. (2013). Human cumulative culture: A comparative perspective. Biological reviews of the Cambridge Philosophical Society. 89. 10.1111/brv.12053.

Derevianko, O. & Shunkov, M. & Kozlikin, Maxim. (2020). Who Were the Denisovans?. Archaeology, Ethnology & Anthropology of Eurasia. 48. 3-32. 10.17746/1563-0110.2020.48.3.003-032.

Dominguez-Rodrigo, Manuel. (2002). Hunting and Scavenging by Early Humans: The State of the Debate. Journal of World Prehistory. 16. 1-54. 10.1023/A:1014507129795.

Dunbar, R. I. M. "The Social Brain: Mind, Language, and Society in

Evolutionary Perspective." *Annual Review of Anthropology*, vol. 32, 2003, pp. 163–81. *JSTOR*, http://www.jstor.org/stable/25064825. Accessed 13 Mar. 2023.

Dunbar, R. I. M. *(1992). "Neocortex size as a constraint on group size in primates". Journal of Human Evolution.* ***22*** *(6): 469–493.* doi:10.1016/0047-2484(92)90081-J

Dunsworth, H.M. Origin of the Genus *Homo* . *Evo Edu Outreach* **3**, 353–366 (2010). https://doi.org/10.1007/s12052-010-0247-8

d'Errico, Francesco & Doyon, Luc & Zhang, Shuangquan & Baumann, Malvina & Galetova, Martina & Gao, Xing & Chen, Fu-You & Zhang, Yue. (2018). The origin and evolution of sewing technologies in Eurasia and North America. Journal of Human Evolution. 125. 71-86. 10.1016/j.jhevol.2018.10.004.

Estévez Escalera, Jordi & Vila, Assumpció. (2013). Analysis of palaeolithic barbed points from the Mediterranean Coast of the Iberian Peninsula: an ethnoarchaeological approach.

Figueirido, Borja & Janis, Christine. (2011). The predatory behaviour of the thylacine: Tasmanian tiger or marsupial wolf?. Biology letters. 7. 937-40. 10.1098/rsbl.2011.0364.

Filatova, Olga & Samarra, Filipa & Deecke, Volker & Ford, John & Miller, Patrick & Yurk, Harald. (2015). Cultural evolution of killer whale calls: Background, mechanisms and consequences. Behaviour. 10.1163/1568539X-00003317.

Forth, Gregory. (2008). Flores after floresiensis: Implications of local reaction to recent palaeoanthropological discoveries on an eastern

Indonesian island. Bijdragen tot de Taal-, Land- en Volkenkunde / Journal of the Humanities and Social Sciences of Southeast Asia and Oceania. 162. 10.1163/22134379-90003670.

Frackowiak, Tomasz & Oleszkiewicz, Anna & Butovskaya, Marina & Groyecka-Bernard, Agata & Karwowski, Maciej & Kowal, Marta & Sorokowski, Piotr. (2020). Subjective Happiness Among Polish and Hadza People. Frontiers in Psychology. 11. 10.3389/fpsyg.2020.01173.

Galetova, Martina. (2019). The symbolism of breast-shaped beads from Dolní Věstonice I (Moravia, Czech Republic). Quaternary International. 503. 221-232. 10.1016/j.quaint.2017.08.035.

Gimenez, Martha E. "Marxism and Class, Gender and Race: Rethinking the Trilogy." Marx, Women, and Capitalist Social Reproduction, Apr. 2018, pp. 82–93., doi:10.1163/9789004291560_005.

Gleeson, Ben. (2019). Human Self-Domestication. 10.1007/978-3-319-16999-6_3856-1.

Gravlee, Clarence C. "How Race Becomes Biology: Embodiment of Social Inequality." American Journal of Physical Anthropology, vol. 139, no. 1, 2009, pp. 47–57., doi:10.1002/ajpa.20983.

Gray, Peter. (2009). Play as a foundation for hunter-gatherer social existence. American Journal of Play. 1. 476-522.

Grouchy P, D'Eleuterio GM, Christiansen MH, Lipson H. On The Evolutionary Origin of Symbolic Communication. Sci Rep. 2016 Oct 10;6:34615. doi: 10.1038/srep34615. PMID: 27721422; PMCID:

PMC5056373.

Groves, Colin. (2007). The Homo floresiensis Controversy. HAYATI Journal of Biosciences. 14. 10.4308/hjb.14.4.123.

Guevara, E. "che." (2003). The motorcycle diaries. Ocean Press

Han, Go. (2016). Hominin interbreeding and the evolution of human variation. Journal of Biological Research. 23. 10.1186/s40709-016-0054-7.

Hauser MD, Yang C, Berwick RC, Tattersall I, Ryan MJ, Watumull J, Chomsky N, Lewontin RC. The mystery of language evolution. Front Psychol. 2014 May 7;5:401. doi: 10.3389/fpsyg.2014.00401. PMID: 24847300; PMCID: PMC4019876.

Hauser, M. D., Chomsky, N., & Fitch, W. T. (2002). The faculty of language: What is it, who has it, and how did it evolve? *Science, 298*(5598), 1569–1579. https://doi.org/10.1126/science.298.5598.1569

Harvati, Katerina. (2007). 100 years of Homo heidelbergensis - Life and times of a controversial taxon. Mitteilungen der Gesellschaft für Urgeschichte, v.16, 85-94 (2007). 16.

Hawkes K, Coxworth JE. Grandmothers and the evolution of human longevity: a review of findings and future directions. Evol Anthropol. 2013 Nov-Dec;22(6):294-302. doi: 10.1002/evan.21382. PMID: 24347503.

Hayashi, Shoji & Kubo, Mugino & Sánchez-Villagra, Marcelo & Taruno, Hiroyuki & Izawa, Masako & Shiroma, Tsunehiro & Nakano, Takayoshi

& Fujita, Masaki. (2020). Variation and mechanisms of life history evolution in insular dwarfism as revealed by a natural experiment. 10.1101/2020.12.23.424186.

Heinrich, Bernd & Bugnyar, Thomas. (2005). Testing Problem Solving in Ravens: String-Pulling to Reach Food. Ethology. 111. 962 - 976. 10.1111/j.1439-0310.2005.01133.x.

Henke, W., Hardt, T. (2011). The Genus *Homo*: Origin, Speciation and Dispersal. In: Condemi, S., Weniger, GC. (eds) Continuity and Discontinuity in the Peopling of Europe. Vertebrate Paleobiology and Paleoanthropology. Springer, Dordrecht. https://doi.org/10.1007/978-94-007-0492-3_3

Henshilwood, Christopher. (2009). The Origins of Symbolism, Spirituality & Shamans: Exploring Middle Stone Age Material Culture in South Africa.

Hicks, James. (2014). The Pre-Symbolic Mind of Homo erectus. 10.13140/RG.2.2.27420.28803.

Hill, Kim & Walker, Robert & Božičević, Miran & Eder, James & Headland, Thomas & Hewlett, Barry & Hurtado, Ana & Marlowe, Frank & Wiessner, Polly & Wood, Brian. (2011). Co-Residence Patterns in Hunter-Gatherer Societies Show Unique Human Social Structure. Science (New York, N.Y.). 331. 1286-9. 10.1126/science.1199071.

Hill, Heather & Weiss, Myriam & Isabelle, Brasseur & Manibusan, Alexander & Sandoval, Irene & Robeck, Todd & Sigman, Julie & Werner, Kristen & Dudzinski, Kathleen. (2022). Killer whale innovation: teaching animals to use their creativity upon request.

Animal Cognition. 25. 1-18. 10.1007/s10071-022-01635-3.

Hoffecker, John. (2005). Innovation and technological knowledge in the Upper Paleolithic of Northern Eurasia. Evolutionary Anthropology: Issues, News, and Reviews. 14. 186 - 198. 10.1002/evan.20066.

Holekamp, Kay & Sawdy, Maggie. (2019). The evolution of matrilineal social systems in fissiped carnivores. Philosophical Transactions of the Royal Society B: Biological Sciences. 374. 20180065. 10.1098/rstb.2018.0065.

Holloway, Ralph. (2009). Cranial Capacity, Neural Reorganization, and Hominid Evolution: A Search for More Suitable Parameters1. American Anthropologist. 68. 103 - 121. 10.1525/aa.1966.68.1.02a00090.

Hovers, Erella, et al. "An Early Case of Color Symbolism: Ochre Use by Modern Humans in Qafzeh Cave." *Current Anthropology*, vol. 44, no. 4, 2003, pp. 491–522. *JSTOR*, https://doi.org/10.1086/375869. Accessed 13 Mar. 2023.

Jensvold, Mary. (2014). Experimental Conversations: Sign Language Studies with Chimpanzees. 10.1007/978-3-319-02669-5_4.

Ji, Qiang & Wu, Wensheng & Ji, Yannan & Li, Qiang & Ni, Xijun. (2021). Late Middle Pleistocene Harbin cranium represents a new Homo species. The Innovation. 2. 100132. 10.1016/j.xinn.2021.100132.

Jiang JH, Rosen PE, Fahy KA. Avoiding the "Great Filter": A Projected Timeframe for Human Expansion Off-World. Galaxies. 2021; 9(3):53. https://doi.org/10.3390/galaxies9030053

Jila, Namu. (2006). Myths and traditional beliefs about the wolf and the crow in Central Asia: Examples from the Turkic Wu-sun and the Mongols. Asian Folklore Studies. 65. 161-177.

Joordens, Josephine & d'Errico, Francesco & Wesselingh, Frank & Munro, Stephen & Vos, John & Wallinga, Jakob & Ankjærgaard, Christina & Reimann, Tony & Wijbrans, Jan R. & Kuiper, K.F. & Mücher, Herman & Coqueugniot, Hélène & Prié, Vincent & Joosten, Ineke & Van Os, Bertil & Schulp, Anne & Panuel, Michel & Haas, Victoria & Lustenhouwer, Wim & Roebroeks, Wil. (2014). Homo Erectus at Trinil on Java Used Shells for Tool Production and Engraving. Nature. advance online publication. 10.1038/nature13962.

Kelly, R. L. (2007). *The Foraging Spectrum: Diversity in Hunter-Gatherer Lifeways*. Eliot Werner Publications. https://doi.org/10.2307/j.ctv2sx9gc9

Key, Catherine A. "The Evolution of Human Life History." *World Archaeology*, vol. 31, no. 3, 2000, pp. 329–50. *JSTOR*, http://www.jstor.org/stable/125105. Accessed 14 Mar. 2023.

Knight, Michael & Van Jaarsveld, Albert & Mills, Michael. (2008). Allo-suckling in spotted hyaenas (Crocuta crocuta): An example of behavioural flexibility in carnivores. African Journal of Ecology. 30. 245 - 251. 10.1111/j.1365-2028.1992.tb00499.x.

Knight, Chris. *Blood Relations: Menstruation and the Origins of Culture.* Yale University Press, 1991. *JSTOR*, http://www.jstor.org/stable/j.ctt5vkr5f. Accessed 14 Mar. 2023.

Kuijper, Bram & Pen, Ido & Weissing, Franz. (2012). A Guide to Sexual Selection Theory. Annu. Rev. Ecol. Evol. Syst.. 43. 287-311. 10.1146/annurev-ecolsys-110411-160245.

Larson, Susan & Jungers, William & Morwood, Michael & Sutikna, Thomas & Jatmiko, & Wahyu, Erlang & Due, Rokus & Djubiantono, Tony. (2008). Homo floresiensis and the evolution of the hominin shoulder. Journal of Human Evolution, 53, 718-731. Journal of human evolution. 53. 718-31. 10.1016/j.jhevol.2007.06.003.

Lewontin, R. C. "The Apportionment of Human Diversity." Evolutionary Biology, 1972, pp. 381–398., doi:10.1007/978-1-4684-9063-3_14.

Lordkipanidze D, Vekua A, Ferring R, Rightmire GP, Agusti J, Kiladze G, Mouskhelishvili A, Nioradze M, Ponce de León MS, Tappen M, Zollikofer CP. Anthropology: the earliest toothless hominin skull. Nature. 2005 Apr 7;434(7034):717-8. doi: 10.1038/434717b. PMID: 15815618.

Malafouris, L. Mind and material engagement. *Phenom Cogn Sci* **18**, 1–17 (2019). https://doi.org/10.1007/s11097-018-9606-7

Malik, Huma & Malik, Fizana. (2022). Emile Durkheim Contributions to Sociology. 6. 7-10.

Marín J, Saladié P, Rodríguez-Hidalgo A, Carbonell E. Neanderthal hunting strategies inferred from mortality profiles within the Abric Romaní sequence. PLoS One. 2017 Nov 22;12(11):e0186970. doi: 10.1371/journal.pone.0186970. PMID: 29166384; PMCID: PMC5699840.

Marx, Karl, 1818-1883. (1996). The Communist manifesto. London ; Chicago, Ill. : Pluto Press,

Marzluff, John & Angell, Tony. (2005). Cultural Coevolution: How the Human Bond with Crows and Ravens Extends Theory and Raises New Questions. Journal of Ecological Anthropology. 9. 69-75. 10.5038/2162-4593.9.1.5.

Mazza, Paul. (2007). Understanding elephant dwarfism on Sicily (Italy) and Flores (Indonesia): Still a long way to go. Human Evolution. 22. 89-95.

Marx, K. (1843) Introduction to a Contribution to the Critique of Hegel's Philosophy of Right. Translated by Mckinnon, A.M. (1976), Collected Works, Vol. 3, New York.

Mcbrearty, Sally. (2003). Patterns of technological change at the origin of Homo sapiens. Before Farming. 2003. 10.3828/bfarm.2003.3.9.

McComb, Karen & Moss, Cynthia & Durant, Sarah & Baker, Lucy & Sayialel, Soila. (2001). Matriarchs As Repositories of Social Knowledge in African Elephants. Science (New York, N.Y.). 292. 491-4. 10.1126/science.1057895.

Meddin, Jay. "Chimpanzees, Symbols, and the Reflective Self." *Social Psychology Quarterly*, vol. 42, no. 2, 1979, pp. 99–109. *JSTOR*, https://doi.org/10.2307/3033691. Accessed 16 Mar. 2023.

Mellars P. Neanderthal symbolism and ornament manufacture: the bursting of a bubble? Proc Natl Acad Sci U S A. 2010 Nov 23;107(47):20147-8. doi: 10.1073/pnas.1014588107. Epub 2010 Nov 15. PMID: 21078972; PMCID: PMC2996706.

Mehr SA, Krasnow MM, Bryant GA, Hagen EH. Toward a productive evolutionary understanding of music. Behav Brain Sci. 2021 Sep 30;44:e122. doi: 10.1017/S0140525X21000030. PMID: 34588071.

Milo, S. (2022). *Why Did Music Evolve? - 4 Hypotheses.* [online] www.youtube.com. Available at: https://www.youtube.com/watch?v=FphCr2My2yg.

Minet, Théo & Deschamps, Marianne & Mangier, Camille & Mourre, Vincent. (2021). Lithic territories during the Late Middle Palaeolithic in the central and western Pyrenees: New data from the Noisetier (Hautes-Pyrénées, France), Gatzarria (Pyrénées-Atlantiques, France) and Abauntz (Navarre, Spain) caves. Journal of Archaeological Science: Reports. 36. 102713. 10.1016/j.jasrep.2020.102713.

Moreau, Luc. (2020). Social inequality before farming? Multidisciplinary approaches to the study of social organization in prehistoric and ethnographic hunter-gatherer-fisher societies. 10.17863/CAM.60627.

Narvaez, Darcia & Gray, Peter & McKenna, James & Fuentes, Agustin & Valentino, Kristin. (2014). Children's Development in Light of Evolution and Culture. 10.1093/acprof:oso/9780199964253.003.0001.

Navarrete, Ana & Isler, Karin & Schaik, Carel. (2009). The Expensive Tissue Hypothesis revisited.

Osipowicz, Grzegorz & Orłowska, Justyna & Piličiauskas, Gytis & Piličiauskienė, Giedrė. (2019). THE STORY OF ONE HARPOON. REUTILIZATION OF AN OSSEOUS PROJECTILE FROM SUBNEOLITHIC AT ŠVENTOJI (LITHUANIA). 29. 237-250. 10.30827/CPAG.v29i0.9775.

O'Hara, Phillip. (2008). "Exploitation and Surplus". The Elgar Companion to Social Economics. 10.4337/9781783478545.00057.

Pearce E, Stringer C, Dunbar RI. New insights into differences in brain organization between Neanderthals and anatomically modern humans. Proc Biol Sci. 2013 Mar 13;280(1758):20130168. doi: 10.1098/rspb.2013.0168. PMID: 23486442; PMCID: PMC3619466.

Peeters S, Zwart H. Neanderthals as familiar strangers and the human spark: How the 'golden years' of Neanderthal research reopen the question of human uniqueness. Hist Philos Life Sci. 2020 Jul 21;42(3):33. doi: 10.1007/s40656-020-00327-w. PMID: 32696095; PMCID: PMC7374475.

Pietarinen, AHTI. (2012). Peirce and Deacon on the Meaning and Evolution of Language. 10.1007/978-94-007-2336-8_4.

Planer and Kim Sterelny, From signal to symbol: The evolution of language. Cambridge, MA: MIT Press, 2021, xx + 272 pp., ISBN: 9780262045971.17 Jun 2022

Pokemon: The Frist Movie: Metwtwo Strikes back. 1998. Takeshi Shudo

Pouyan, Nasser. (2016). MESOPOTAMIA, THE CRADLE OF CIVILIZATION AND MEDICINE. World Journal of Pharmaceutical Research. 5. 192-225. 10.20959/wjpr20164-5895.

Ur, J. (2014). Households and the Emergence of Cities in Ancient Mesopotamia. Cambridge Archaeological Journal, 24(2), 249-268. doi:10.1017/S095977431400047X

Rendell, Luke & Whitehead, Hal. (2001). Culture in Whales and Dolphins. The Behavioral and brain sciences. 24. 309-24; discussion 324. 10.1017/S0140525X0100396X.

Richerson, Peter & Boyd, Robert. (1999). The Evolution of Human Ultra-sociality.

Richter JN, Hochner B, Kuba MJ (2016) Pull or Push? Octopuses Solve a Puzzle Problem. PLOS ONE 11(3): e0152048. https://doi.org/10.1371/

journal.pone.0152048

Rightmire, G.. (2004). Brain Size and Encephalization in Early to Mid-Pleistocene Homo. American journal of physical anthropology. 124. 109-23. 10.1002/ajpa.10346.

Rocca, Roxane & Connet, Nelly & Lhomme, Vincent. (2017). Before the transition? The final middle Palaeolithic lithic industry from the Grotte du Renne (layer XI) at Arcy-sur-Cure (Burgundy, France). Comptes Rendus Palevol. 16. 878–893. 10.1016/j.crpv.2017.04.003.

Rolando, Antonio & Zunino, Mario. (1992). Observations of Tool Use in Corvids. Ornis Scandinavica. 23. 201-202. 10.2307/3676451.

Røskaft, Eivin, and Yngve Espmark. "Vocal Communication by the Rook Corvus Frugilegus during the Breeding Season." *Ornis Scandinavica (Scandinavian Journal of Ornithology)*, vol. 13, no. 1, 1982, pp. 38–46.*JSTOR*, https://doi.org/10.2307/3675971. Accessed 13 Mar. 2023.

Rousseau, Jean-Jacques, 1712-1778. On the Social Contract ; Discourse on the Origin of Inequality ; Discourse on Political Economy. Indianapolis :Hackett Pub. Co., 1983.

Ruff, Christopher. (2005). Climatic adaptation and hominid evolution: The thermoregulatory imperative. Evolutionary Anthropology: Issues, News, and Reviews. 2. 53 - 60. 10.1002/evan.1360020207.

Ryan, Alan. (2012). The Nature of Human Nature in Hobbes and Rousseau. 10.23943/princeton/9780691148403.003.0012.

Savage PE, Loui P, Tarr B, Schachner A, Glowacki L, Mithen S, Fitch WT. Music as a coevolved system for social bonding. Behav Brain Sci. 2020 Aug 20;44:e59. doi: 10.1017/S0140525X20000333. PMID: 32814608.

Schoch, Werner H. & Bigga, Gerlinde & Böhner, Utz & Richter, Pascale & Terberger, Thomas. (2015). New insights on the wooden weapons from the Paleolithic site of Schöningen. Journal of human evolution. 89. 10.1016/j.jhevol.2015.08.004.

Schmandt-Besserat, Denise. "The Earliest Precursor of Writing." *Scientific American*, vol. 238, no. 6, 1978, pp. 50–59. *JSTOR*, http://www.jstor.org/stable/24955753. Accessed 16 Mar. 2023.

Serpell, James. (2013). Domestication and history of the cat. 10.1017/CBO9781139177177.011.

Shapiro, B & Leducq, Jean-Baptiste & Mallet, James. (2016). What Is Speciation?. PLoS genetics. 12. e1005860. 10.1371/journal.pgen.1005860.

Shaver, John & Bulbulia, Joseph. (2016). Signaling Theory and Religion.

Shultz S, Nelson E, Dunbar RI. Hominin cognitive evolution: identifying patterns and processes in the fossil and archaeological record. Philos Trans R Soc Lond B Biol Sci. 2012 Aug 5;367(1599):2130-40. doi: 10.1098/rstb.2012.0115. PMID: 22734056; PMCID: PMC3385680.

Shipton C, Clarkson C, Bernal MA, Boivin N, Finlayson C, Finlayson G, Fa D, Pacheco FG, Petraglia M. Variation in lithic technological strategies among the Neanderthals of Gibraltar. PLoS One. 2013 Jun 6;8(6):e65185. doi: 10.1371/journal.pone.0065185. PMID: 23762312; PMCID: PMC3675147.

Short, T. (2007). The Development of Peirce's Semeiotic. In Peirce's Theory of Signs (pp. 27-59). Cambridge: Cambridge University Press. doi:10.1017/CBO9780511498350.003

Smith, K. Evolution of a single gene linked to language. *Nature* (2009). https://doi.org/10.1038/news.2009.1079

Spikins, Penny & Needham, Andy & Tilley, Lorna & Hitchens, Gail. (2018). Open Access: Calculated or caring? Neanderthal healthcare in social context. 10.4324/9781003164623-2.

Spikins, Penny. (2018). How Compassion Made Us Human: The origins of tenderness, trust and morality.

Spikins, Penny. “Goodwill Hunting? Debates over the ‘meaning’ of Lower Palaeolithic Handaxe Form Revisited.” *World Archaeology*, vol. 44, no. 3, 2012, pp. 378–92. *JSTOR*, http://www.jstor.org/stable/42003537. Accessed 14 Mar. 2023.

Steegmann, A & Cerny, Frank & Holliday, Trenton. (2002).

Neandertal cold adaptation: Physiological and energetic factors. American journal of human biology : the official journal of the Human Biology Council. 14. 566-83. 10.1002/ajhb.10070.

Stringer CB, Barnes I. Deciphering the Denisovans. Proc Natl Acad Sci U S A. 2015 Dec 22;112(51):15542-3. doi: 10.1073/pnas.1522477112. Epub 2015 Dec 14. PMID: 26668361; PMCID: PMC4697382.

Stringer, Christopher. (2012). The status of Homo heidelbergensis (Schoetensack 1908). Evolutionary anthropology. 21. 101-7. 10.1002/evan.21311.

Swift K, Marzluff JM. Occurrence and variability of tactile interactions between wild American crows and dead conspecifics. Philos Trans R Soc Lond B Biol Sci. 2018 Sep 5;373(1754):20170259. doi: 10.1098/rstb.2017.0259. PMID: 30012745; PMCID: PMC6053988.

Számadó S, Szathmáry E. Language Evolution. PLoS Biol. 2004 Oct;2(10):e346. doi: 10.1371/journal.pbio.0020346. Epub 2004 Oct 12. PMCID: PMC521730.

The Independent. (2016). *Amazon tribeswomen escape back to forest after rejecting civilisation.* [online] Available at: https://www.independent.co.uk/news/world/americas/amazon-tribeswomen-escape-back-to-forest-after-rejecting-civilisation-a7237796.html [Accessed 14 Mar. 2023].

Theofanopoulou C, Gastaldon S, O'Rourke T, Samuels BD, Messner A, et al. (2017) Self-domestication in *Homo sapiens*: Insights from comparative genomics. PLOS ONE 12(10): e0185306. https://doi.org/10.1371/journal.pone.0185306

Theis, Wolfgang. (2010). Language, Culture and Symbolic Forms.

"Tigers Evolved with Snow Leopards, Gene Study Reveals." Big Cat Rescue, 18 Feb. 2013, bigcatrescue.org/tigers-evolved-with-snow-leopards-gene-study-reveals/.

Tostevin, Gilbert. (2000). Behavioral change and regional variation across the Middle to Upper Paleolithic transition in Central Europe, Eastern Europe, and the Levant /.

Trivers, Robert. (1974). Parent-Offspring Conflict. 14. 249-264.

van der Geer, Alexandra & Lyras, George & Vos, John. (2021). The Island Rule: Dwarfism and Gigantism. 10.1002/9781119675754.ch23.

Verdún, Ester & Casabo, Josep. (2021). Shellfish consumption in the Early Upper Palaeolithic on the Mediterranean coast of the Iberian Peninsula: The example of Foradada Cave.

Villier, Boris & van den Hoek Ostende, Lars & Vos, John & Pavia, Marco. (2013). New discoveries on the giant hedgehog Deinogalerix from the Miocene of Gargano (Apulia, Italy). Geobios. 46. 63-75. 10.1016/j.geobios.2012.10.001.

Von Petzinger, G. (2017). *The first signs : unlocking the mysteries of the world's oldest symbols.* First Atria Paperback edition. New York: Atria Paperback.

Voormolen, Boudewijn. (2008). Ancient hunters, modern butchers: Schöningen 13II - 4, a kill-butchery site dating from the northwest European Lower Palaeolithic. Journal of Taphonomy. 6.

Wales, Nathan. (2012). Modeling Neanderthal clothing using ethnographic analogues. Journal of human evolution. 63. 10.1016/j.jhevol.2012.08.006.

de Waal, F. (2006). *Primates and philosophers: How morality evolved.* (S. Macedo & J. Ober, Eds.). Princeton University Press.

Watts, Ian. (1992). Chris Knight. 1991. Blood Relations: Menstruation and the Origins of Culture.. Papers from the Institute of Archaeology. 3. 84. 10.5334/pia.26.

Watts, Ian. "Ochre in the Middle Stone Age of Southern Africa: Ritualised Display or Hide Preservative?" *The South African Archaeological Bulletin*, vol. 57, no. 175, 2002, pp. 1–14. *JSTOR*, https://doi.org/10.2307/3889102. Accessed 16 Mar. 2023.

Whitehead, Hal. (2015). Life History Evolution: What Does a Menopausal Killer Whale Do?. Current Biology. 25. 10.1016/j.cub.2015.02.002.

Wiessner, Polly. (2014). Embers of Society: Firelight Talk among the Ju/'hoansi Bushmen. Proceedings of the National Academy of Sciences of the United States of America. 111. 10.1073/pnas.1404212111.

Weiss, Kenneth. (2010). "Nature, Red in Tooth and Claw", So What?. Evolutionary Anthropology - EVOL ANTHROPOL. 19. 41-45. 10.1002/evan.20255.

Wengrow, David & Graeber, David. (2015). Farewell to the 'childhood of man': Ritual, seasonality, and the origins of inequality. Journal of the Royal Anthropological Institute. 21. 10.1111/1467-9655.12247.

Wilson, Michael & Wallauer, William & Pusey, Anne. (2004). New Cases of Intergroup Violence Among Chimpanzees in Gombe National Park, Tanzania. International Journal of Primatology. 25. 523-549. 10.1023/B:IJOP.0000023574.38219.92.

Wrangham, Richard. (2017). Control of Fire in the Paleolithic: Evaluating the Cooking Hypothesis. Current Anthropology. 58. S000-S000. 10.1086/692113.

Wu, Lilian Shiao-Yen, and Daniel B. Botkin. "Of Elephants and Men: A Discrete, Stochastic Model for Long-Lived Species with Complex Life Histories." *The American Naturalist*, vol. 116, no. 6, 1980, pp. 831–49. *JSTOR*, http://www.jstor.org/stable/2460410. Accessed 14 Mar. 2023.

Young, J. Z. "The Buccal Nervous System of Octopus." *Philosophical Transactions of the Royal Society of London. Series B, Biological Sciences*, vol. 249, no. 755, 1965, pp. 27–44. *JSTOR*, http://www.jstor.org/stable/2416591. Accessed 13 Mar. 2023.

Yravedra, J., Solano, J.A., Courtenay, L.A. *et al.* Use of meat resources in the Early Pleistocene assemblages from Fuente Nueva 3 (Orce, Granada, Spain). *Archaeol Anthropol Sci* **13**, 213 (2021). https://doi.org/10.1007/s12520-021-01461-7

Zahavi, Amotz. (1997). The Handicap Principle: A Missing Piece of

Darwin's Puzzle.

Zipes, Jack. "The Meaning of Fairy Tale within the Evolution of Culture." *Marvels & Tales*, vol. 25, no. 2, 2011, pp. 221–43. *JSTOR*, http://www.jstor.org/stable/41389000. Accessed 16 Mar. 2023.

Zuiddam, Benno. (2019). Was evolution invented by Greek philosophers?. JOURNAL OF CREATION. 68-75.

Printed in Great Britain
by Amazon

36102165R00059